TRANSACTIONS

OF THE

AMERICAN PHILOSOPHICAL SOCIETY

HELD AT PHILADELPHIA
FOR PROMOTING USEFUL KNOWLEDGE

NEW SERIES—VOLUME XXXI, PART III

DECEMBER, 1940

PHYTOPLANKTON AND PLANKTONIC PROTOZOA OF THE OFFSHORE WATERS GULF OF MAINE

HENRY B. BIGELOW, LOIS C. LILLICK AND MARY SEARS

PHILADELPHIA:

THE AMERICAN PHILOSOPHICAL SOCIETY

104 SOUTH FIFTH STREET

1940

LANCASTER PRESS, INC., LANCASTER, PA.

PHYTOPLANKTON AND PLANKTONIC PROTOZOA OF THE OFFSHORE WATERS OF THE GULF OF MAINE

Part I—Numerical Distribution [1]

By Henry B. Bigelow, Lois C. Lillick, and Mary Sears

CONTENTS

ABSTRACT

Except for occasional local flowerings, the phytoplankton is scanty from late autumn until early spring (av. 1,852,000–7,856,000 cells per column). In the one year of record the minimum was reached in December in the western coastal belt and over George's Bank, but not until January in the eastern side generally. In most years, an increase in the phytoplankton takes place by mid- or late March (late February in an early spring) in the western coastal belt, western George's Bank and in Passamoquoddy Bay. In some years, the flowering may start equally early in the eastern side of the Gulf, south from Nova Scotia and over the eastern part of George's Bank, but not until April or even May in the northern coastal belt or over the bowl generally.

[1] Contribution No. 259 from the Woods Hole Oceanographic Institution.

The vernal climax (averaging 2–14 billion cells per column) is reached by early or mid-April in the vicinity of Massachusetts Bay and over the eastern part of the Gulf generally, by late April in the southern part of the western basin, shortly thereafter in the northern coastal belt, and latest (May or June) in the Bay of Fundy region.

Subsequent to the vernal climax, a rich flora may persist in the northern coastal belt, in the Bay of Fundy and (judging from available evidence) also on George's Bank through June or July. Elsewhere an abrupt impoverishment follows the vernal climax—so severe in the open basin as to make the May counts lower than at any other season.

A second flowering takes place through August, the more productive centers being in the Bay of Fundy region and on George's Bank. Similar centers occur in the Massachusetts Bay region and southward in September. This increase is followed by a progressive autumnal impoverishment whereby the winter condition is re-established.

The initiation of the vernal flowerings is apparently not dependent on the available nutrients, on light, or on the attainment of any particular temperature, but is possibly conditioned by a certain amount of stability in the water column. Nutrients, on the other hand, play a part in the termination of the vernal multiplication, nitrates perhaps being more likely to act as the limiting factor than phosphates. Apparently, the nitrate supply is replenished in the illuminated zone before the second flowering.

The planktonic protozoa of the Gulf of Maine, at their lowest level from October on, increase considerably from May through June, reaching a maximum in July.

INTRODUCTION

The general composition of the plankton of the Gulf of Maine, both animal and vegetable,[2] is now well known. Davidson (1934) and Gran and Braarud (1935) have also contributed quantitative studies of the latter for the region of the Bay of Fundy. The collections [3] made on the periodic cruises of "Atlantis" during the year 1933–1934 now provide the basis for a general survey of the numerical distribution of planktonic plant cells for the gulf as a whole, from season to season.

Sources of Information

Quantitative collections at representative localities (Figs. 2, 3, 4) made from "Atlantis" in June, July, September, October and December, 1933, and in January, March, April, May, June and July, 1934 and in August, 1936 form the chief basis of the present report. Information has also been freely drawn from earlier publications, both for the offshore waters of the gulf (Bigelow, 1914–1922, 1926; Gran, 1933; Braarud, 1934; Gran and Braarud, 1935), for the Bay of Fundy, and for various inshore localities (Bailey, 1910–1924; Bailey and Mackay, 1921; McMurrich, 1917; Fritz, 1921; Davidson, 1934; Burkholder, 1933; Gran and Braarud, 1935).

Methods

The first efforts at quantitative determination for the general region were based on horizontal net-hauls (Fritz, 1921; Bigelow, 1926; Davidson, 1934). Lacking information as to the amount of water strained, these not only involved all the errors of the tow-net method *per se*, but could yield only relative information at best. More recent studies, however, have been based on determinations of amounts of plankton contained in known volumes of water. And successive investigators can claim increasingly greater accuracy, for

[2] See especially Bigelow, 1926.

[3] The original data are on file at the Woods Hole Oceanographic Institution. For a preliminary account see Lillick (1938).

whereas Burkholder (1933) filtered the water sample through fine bolting silk, Gran (1933), Braarud (1934) and Gran and Braarud (1935) employed the "centrifuge" method, and the counts of "Atlantis" samples here presented were by Untermoehl's (1931) precipitation method as modified by Steemann Nielsen and Von Brand (1934).

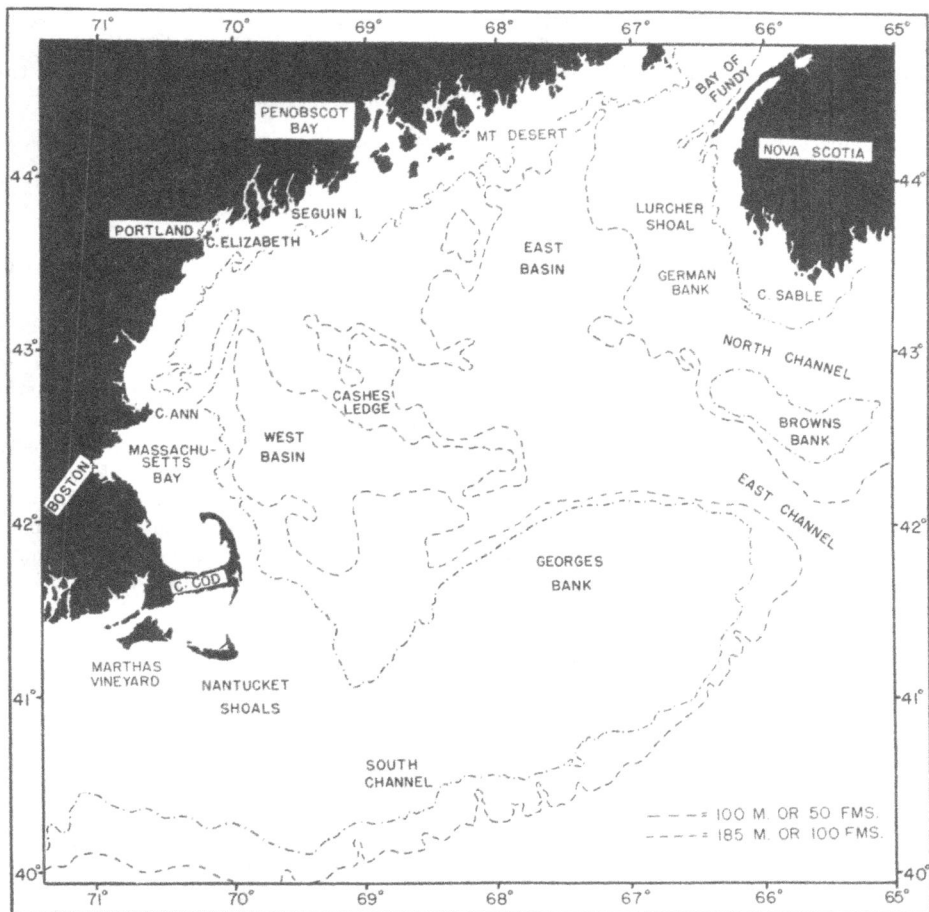

FIG. 1.   Chart of the Gulf of Maine showing localities mentioned in the text.

Our standard procedure was to make counts of 20–50 cc. taken from a 1-liter sample of water collected with the Nansen water bottle at depths of 1, 10, 30, 50 and 80 meters, 10% being added to the counts to compensate for the amount lost during the operation as Gran (1933, p. 160) has shown to be necessary.   Most, however, of the counts for the

April–May and May–June cruises of 1934 were of aliquot samples (concentrated over a membrane filter) of catches made in vertical tows from 100 meters (or less) with ordinary tow-nets of No. 20 silk, 35 cm. in diameter.

Our aim has been to present a picture of the numbers of plant cells existing in the water column as a whole at different seasons and localities, as one step in current studies of organic production in the Gulf. The numbers of cells are therefore stated per 0.1 square meter of sea surface rather than per liter. In the case of the water-bottle collections, this value is arrived at by the following arithmetic calculation, the assumption being that practically the whole phytoplankton was in the upper 80 meters:

1. Summation of (Σ) (average number of cells per liter
$$\frac{\times \text{ depth in meters of each stratum sampled)}}{\text{Total depth sampled in meters}} = \text{Average number of cells per liter.}$$

2. Average number of cells per liter $\times$ number of liters
under 0.1 square meter to depth sampled $= \text{Total number of cells under 0.1 square meter.}$

In the case of the vertical hauls the catches should approximate the desired value directly (since the nets were practically 0.1 square meter in diameter) were it not for, (*a*) the failure of even the finest meshed silk to retain the smallest cells, and (*b*) the reduction in straining efficiency that results as the meshes become clogged. In the present case the parallel tests tabulated below have shown that the total catches were seriously affected by both these factors, and especially by clogging. In rich areas in fact, the nets caught on the average only about 1% as many cells per column as did the water bottles; about 2% as many in areas where the flora was somewhat less dense; and not over about 10% as many in regions of sparsity. To render the net counts comparable with those of the volumetric samples, those of 10 thousand to 5 million per 0.1 square meter were therefore multiplied by 10; those of 5 to 9 million by 50; and those over 9 million by 100.

TOTAL NUMBER OF CELLS PER COLUMN

Location of Station	Net Catch		Precipitation Method
	Uncorrected	Corrected	
Off Cape Elizabeth..............	8,940,000	447,000,000	504,000,000
Western Basin (north)..........	152,329	1,523,000	970,000
Off Mt. Desert Island...........	32,388,600	3,238,800,000	2,000,000,000
Off western Nova Scotia.........	27,007,750	2,700,000,000	1,700,000,000
Eastern Basin...................	53,162,000	5,316,000,000	5,375,000,000
Eastern Basin...................	21,213,360	2,121,336,000	2,176,500,000
Northern Channel...............	9,199,280	919,928,000	1,778,000,000
Eastern Channel................	38,250	380,000	1,700,000
Southern Channel...............	29,260	290,000	852,000
Off northern tip of Cape Cod.....	73,000	730,000	900,000
Western Basin (south)..........	1,125,000	11,250,000	13,000,000
Off Cape Elizabeth..............	6,750	67,500	192,000
Western Basin (north)..........	92,000	920,000	1,000,000

RELIABILITY OF RESULTS

In evaluating the reliability of results drawn from data of the sorts just listed, the adequacy of the regional sampling, i.e. whether the stations were sufficiently close together, is perhaps the most important question, it being well known that the distribution of the phytoplankton may be very irregular with sharp contrasts in abundance within short distances, and especially so during periods of active multiplication. Instances for the Gulf of Maine are given on page 160. Hence the criticism might be made that stations as far apart as ours might fail to reveal such conditions. In the Gulf, however, a marked streakiness occurs most often in shoal water, the plankton being much more uniformly distributed in the deeper bowl outside the 100-meter contour, where it may in fact be essentially uniform over large areas (Bigelow, 1926, p. 403). The uniformity of the gradation from regions of high counts to those of low for the open Gulf shown on the chart for the month when counts were highest (Fig. 2) and when, consequently, the greatest streakiness was to be expected, with the generally close correspondence between the counts for 1933–1934 and those earlier recorded, appears to us to justify the conclusion that the grid of stations was sufficiently close in that year to reveal at least the major variations in the phytoplankton, regional as well as secular.

Errors inherent in the calculations of numbers probably average about 10% for the water-bottle collections (Steemann Nielsen, 1938), at least for such fractions of the flora as the method yielded.[4] The error may, however, be much wider in individual cases for the net hauls, not only because of the shortcomings inherent in this method of collection now generally recognized, or because of uncertainty as to the precise length of the columns of water fished, resulting from deviations from the vertical due to the drift of the ship, but especially because of the incalculable error introduced in our attempt to adjust the results of the net catches to correspond with those of the water-bottles, as stated above. The fact, however, that counts per 0.1 square meter for the northern part of the gulf for the spring of 1932, calculated from Gran and Braarud's (1935) table, correspond closely with our adjusted counts at neighboring localities at that same season (of the spring maximum) in 1934, suggest that the general picture yielded by these is essentially correct. Furthermore, since the probable error varies inversely as the number of organisms per sample, it should be least during the spring maximum, and even if the error in any individual case were as great as 1000%, of which there is small liklihood, the final picture would be little altered.

NUMERICAL OCCURRENCE OF PHYTOPLANKTON (DIATOMS, PERIDINIANS, COCCOLITHOPHORIDS AND SILICOFLAGELLATES)

We may anticipate the second part of this report (Lillick, 1940) by pointing out that the data for 1934 corroborate the joint evidence of earlier years, to the effect that in the Gulf of Maine the major fluctuations in the total number of planktonic plant cells in the water, regional and seasonal, are due chiefly to fluctuations in the abundance of diatoms, with peridinians, coccoliths and other groups playing only a secondary rôle.

---

[4] One criticism of the precipitation method for taxonomic investigations is that many of the more delicate organisms are distorted beyond recognition.

Winter Minimum

Previous observations had already shown that in the Gulf of Maine, as in boreal seas generally, inshore as well as offshore, the planktonic flora (chiefly of diatoms) is not only sparse from late autumn through the winter, but also that it is the most nearly constant in abundance at that time of year.   In Massachusetts Bay, for example, the flora proved very scanty throughout the winter of 1912–1913, contrasted with the abundant flowering that took place early in the subsequent spring (Bigelow, 1914a).   Fritz's (1921) and Davidson's

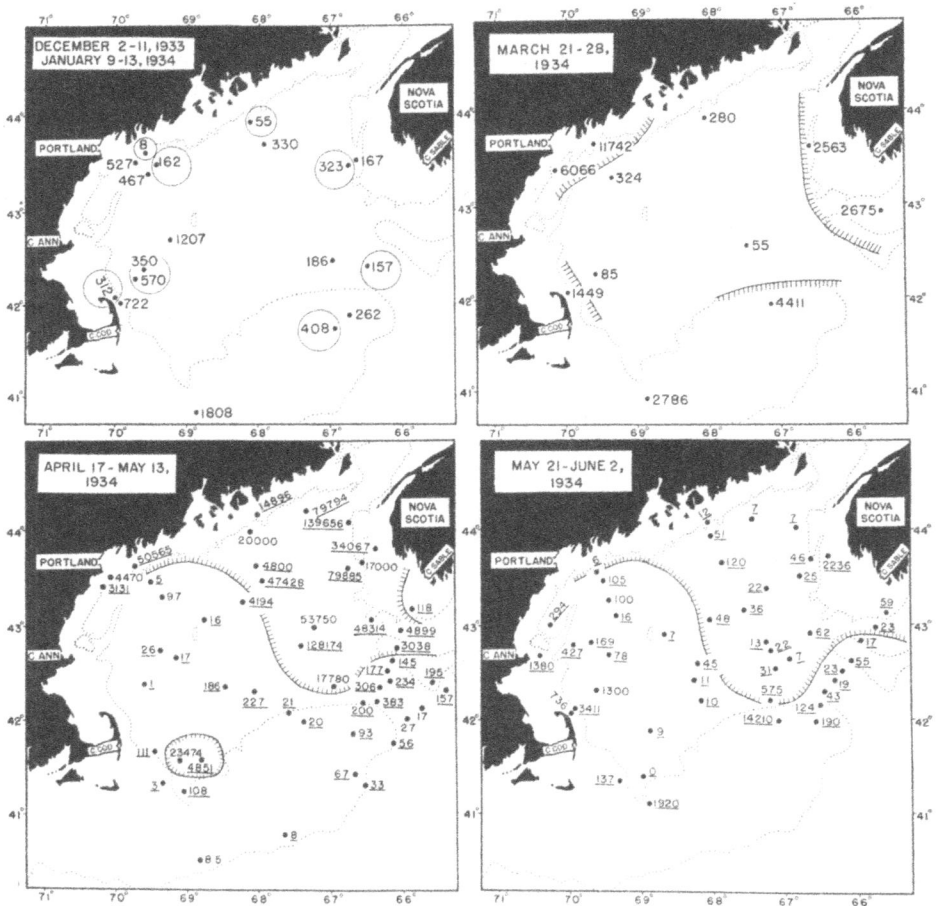

Fig. 2. Numbers of phytoplankton cells, in tens of thousands, for December (encircled), January and March, and May–June; and in hundreds of thousands for April–May, 1933–1934, per 0.1 square meter of sea surface. Counts based on vertical hauls are underlined. Hatched boundaries enclose areas where greatest change took place since the preceding cruise.

(1934) more recent towings in the shallow waters of Passamoquoddy Bay tributary to the Bay of Fundy yielded similar results, namely much smaller quantities generally in early and midwinter than at other seasons in each year of record.   When opportunity came in 1920–1921 to reconnoitre the gulf as a whole, it proved that these localities in its opposite sides typify not only the coastwise belt as a whole in this respect, but in all probability the shoal waters of George's Bank as well (Bigelow, 1926).   In the deep central basin, however, the absolute yearly minimum falls in late spring after the eclipse of the vernal outburst (p. 161).

The water-bottle samples for 1933–1934 corroborate this picture of winter scarcity, since the counts for December and January combined average only 4,171,000 cells per column (18 stations, Fig. 2),[5] contrasted with an average of more than one billion per column in late April and early May (Fig. 2).   In the bowl of the gulf generally, outside the 100-meter contour, averages of 1,500,000 for December (5 stations) and 5,081,000 for January (5 stations), but 1,863,000 only for March (4 stations) suggest some slight increase in production for midwinter, followed by a decline to approximately the early winter level (see also Part II of this report).   In the western coastal belt, however, counts at two stations of 80,000 and of 3,120,000 in December, but of 5,269,000 and 7,216,000 in January, and of 117,425,000 and 14,493,000 in March, point to a definite early winter minimum, followed by a slight but unmistakeable increase in plant cells through the late winter. This appears to apply equally to the waters southward across the southern rim of the gulf, judging from counts there of 18 million in January and 28 million in March (no data for December).   Local flowerings may, however, develop in shoal water close to land even in early midwinter, witness the record in 1925 of "a rich flowering of *Rhizosolenia alata* from the middle of December (appearing between the 10th and 15th) through January" in Cape Cod Bay (Bigelow, 1926, p. 396) and of Chaetoceros in abundance in Ipswich Bay, a few miles north of Cape Ann, on January 30, 1913 (Bigelow, 1914a, p. 405; 1926, p. 395).   The fact that so short a record of observation should have revealed these flowerings is evidence that temporary enrichment of the water with planktonic vegetation is not an exceptional event in shoal water even at that season, when phytoplankton is usually scarce.   In the eastern coastal waters and on the eastern part of George's Bank the few available counts were only slightly more than half as large for January (2 million to 3 million) as for December (3 million to 4 million), though again considerably larger for March (26 million to 44 million).   Davidson's (1934, Fig. 4) graph of average monthly fluctuations in diatoms similarly shows the minimum as falling in January or February more often (4 years) than in November or December (2 years).

If combinations of the years of record can be taken as representative, it thus appears that the minimum falls a month or so earlier, and is sooner succeeded by an appreciable enrichment of the flora, in the western coastal belt and thence southward than in the basin generally, in the eastern coastal belt, or on the eastern part of George's Bank.   Unfortunately, we lack pertinent data in this respect for the northern coastal belt, but if Passomoquoddy Bay can be taken as representative of it, as seems reasonable, it stands intermediate between the two extremes just outlined.   By Davidson's (1934) data, however, the precise

[5] The observations are so few in number for December and for January that the combination of the data for the two months probably gives a more reliable picture than would those for either one if taken separately.

date when the winter flora is the most scanty may be expected to vary by as much as three months from year to year at any given locality in inshore waters—from early December to February in Passamoquoddy Bay where her towings were made. We have no information as to how the range of annual variation compares with this in the offshore waters of the gulf.

We may note in passing that the lowest winter count recorded in 1933–1934 was 80,000 per column (Fig. 2), the highest about 18 million (Fig. 2), the next highest about 7 million.

VERNAL FLOWERINGS

The early and midwinter flowerings noted above (p. 155) for the western coastal belt appear to have been sporadic events, since the diatoms concerned were not those responsible for the vernal outburst,[6] and it is certain that the more productive of the two was restricted to a small area, as described elsewhere (Bigelow, 1926, p. 396). A significant increase in the number of vegetable cells took place, however, in Massachusetts Bay (representative of the western coastal belt) in 1913 between February 13th and the 4th of March (Bigelow, 1914a, p. 408). In 1920 (Bigelow, 1926, Fig. 108) there were considerable quantities (up to 290 cc. per haul) of diatoms by March 4 in the coastal belt off Portland, Maine, where (by other evidence) no richer flora is to be expected in January or February than elsewhere around the coast line of the gulf. And again in 1934, the counts were significantly larger for the last week in March than for January at each of the three stations in the western coastal belt where data were obtained in both these months (Fig. 2).

Fritz (1921) likewise records considerably larger catches for March than for January–February in Passamoquoddy Bay [7] for the year 1916, and Davidson (1934) described the multiplication of diatoms as commencing there as early as February in one year (1931), and in March in four years, though not until April in one (1925), for the periods 1925–1928 and 1930–1931.[8] As a result, perhaps, of the discharge from Passamoquoddy Bay, Davidson's (1934) curves similarly show a rapid increase (of diatoms) in March at her station (No. 5) just off its mouth in each of the five years (1926–1928; 1930–1931), consistent with counts by Gran and Braarud (1935) corresponding to about 12,393,000 cells per column at this same general locality on March 5 of 1932.[9] It appears that vernal enrichment does not involve the waters of the open Bay of Fundy generally until somewhat later in the season, for Gran and Braarud describe the phytoplankton there as still continuing very scanty in March. The limits of the richer and poorer subdivisions of the Bay were, in fact, rather precise at the time though somewhat spotty (Gran and Braarud, 1935, p. 333, Fig. 35).

We may conclude from the foregoing that active multiplication of the phytoplankton is ordinarily in progress in the western coastal belt on the one hand, and in the waters tributary to the Bay of Fundy on the other, by early March or by the end of the month at latest. The evidence of 1920 points to the general vicinity of Cape Elizabeth as the site of the earliest outburst of all, since on the fourth of the month in that year the tow yielded

[6] See Part II of this report (Lillick, 1940) for description of the qualitative composition of the flora from season to season.

[7] Fritz (1921, pp. 50–52) reports averages of 13,200 cells per tow for February (2 stations) but of 112,525 per tow for March (4 stations).

[8] Data for the critical season are lacking for 1929.

[9] Average of 1549 per liter for the several sampling levels combined.

290 cc. there, but only 5–8 cc. off Massachusetts Bay to the south and 2 cc. near Mt. Desert Island to the east, a distribution which is at least consistent with the fact that a considerably larger count was recorded off Cape Elizabeth than near Cape Cod in late March of 1934 (Fig. 2). The few available data (for 1920 and 1934) suggest that flowerings may be expected to develop equally early on the western part of George's Bank.

Lacking coincident data both for the Massachusetts Bay region and for Passamoquoddy Bay in any one year, we have no basis for judging whether the difference in latitude between these two localities is accompanied by a corresponding difference in the date of inception of active vernal flowering. Actually this event has been recorded as early (February) in the more northern of these two localities in the year of record when it fell earliest there (1931), as in the more southern (late February in 1925, see above). An appreciable increase in the phytoplankton may also occur equally early in some years along the eastern coastal belt to the southward of Nova Scotia and on the eastern part of George's Bank, as happened in 1934 when (at the pertinent stations) the count was between twelve and thirteen times as great there in the last week in March (26–44 million) as it had been two months previously (about 2–3 million, Fig. 2). However, the fact that the phytoplankton still continued minimal in this same general region as late as the 23rd of March in 1920 (Bigelow, 1926, Fig. 108) is evidence that a variation of a month or more is to be expected there in the spring outburst from year to year. The small counts recorded for the bowl of the gulf generally outside the 200-meter contour for late March of 1934 (from 550,000 to about 3 million cells per column, Fig. 2), added to the correspondingly small volumes yielded there by the towings in this same month of 1920 (Bigelow, 1926, Fig. 108), make it unlikely that active flowerings ever develop anywhere in that general subdivision of the gulf before April in normal years. Furthermore it may prove that this applies equally to the northern coastal belt between Penobscot Bay and the Bay of Fundy, since the volumes of phytoplankton were still minimal there at 2 stations on March 3 and March 22 in 1920 (Bigelow, 1926, Fig. 108); but no data are available for that particular sector at the critical season in any other year.

The regional and annual variations outlined above, added to the lack of data at critical localities and dates, make it difficult to present a coherent picture of the development of the vernal flowerings which can be taken as characteristic of a normal year. It is, however, clear that the peak may be reached at the mouth of Massachusetts Bay as early as the end of March in some years, as happened in 1938 and again in 1939, when a great abundance of diatoms in the last half of the month was followed by a rapid decline by the first week in April (unpublished data). In other years, on the other hand, great numbers of diatoms may persist there as late as the third week of April; in 1920, for example, the catch was as large there (200 cc.) on April 20th as it had been on the 6th (190 cc.). Lacking pertinent data for 1934, the first week in April might be set as the normal expectation for this point in the seasonal cycle in this particular region.

Cape Ann appears, however, to mark a natural division in this respect, because in 1913 diatoms continued extremely abundant near the Isles of Shoals some 20 miles north of the cape until at least the first week in May (Bigelow, 1914a), i.e. until some 2–3 weeks later than is likely to be the case just south of the cape. Again in 1915 diatoms were swarming all along the coast inshore as well as offshore from the vicinity of Portland to the Grand Manan Channel as late as May 10–13 (Bigelow, 1917, p. 324, Fig. 98). Similarly in 1934,

5 billion cells per column were recorded off Portland as late as the third week in April, and 50–60 million per column near Cape Elizabeth, while the fact that the nitrates and phosphates had declined at these stations (on the average) to 5–20 mgs. and 34–36 mgs., respectively, from 100 ± mgs. of nitrate and 100 ± mgs. (at 1 meter) of phosphate in March, suggests that the actual peak of phytoplankton abundance had passed some days earlier. Counts of 1–3 billion were also recorded in the coastal waters near Mt. Desert Island in that same month of 1932,[10] with 356 million cells per column on April 29.   In Passamoquoddy Bay, Fritz's (1921), Davidson's (1934), and Gran and Braarud's (1935) counts show the vernal outburst as reaching its climax rarely as early as April (about 4 billion cells per column at one station on April 25, 1932, as calculated from Gran and Braarud's counts), but usually not until May.   Available data suggest that the climax is not ordinarily reached in the open Bay of Fundy until several weeks later, for Davidson's (1934) curves for her station off the mouth of Passamoquoddy Bay show the peak as falling in June in four years of her series, and late in May in a fifth.   Gran and Braarud (1935, p. 339), however, describe an "extreme decline in phytoplankton" from May to June for the Bay of Fundy in a year (1932) when a rich flora (though perhaps not the extreme peak) had developed in Passamoquoddy Bay in April, as remarked above.

Continuing around the Gulf, the data for 1934 show a great abundance of phytoplankton in the last half of April throughout the northeast part generally, including both the coastal waters off western Nova Scotia, and the eastern part of the basin (Fig. 2); locally also off southern Nova Scotia and out to Brown's Bank (though interspersed with relatively barren localities) early in that May.

Unfortunately the data do not serve accurately to trace the development of the spring flowering for any part of George's Bank or for the western arm of the basin of the Gulf. Pertinent records for the western part of the bank are, (a) a catch of 200 cc. (moderately rich) on February 22, 1922 (Bigelow, 1926, p. 399, Fig. 108); (b) a rich diatom plankton there on April 26–27, 1913 (Bigelow, 1914a, p. 415); and (c) counts of 28 million in March (1 station, Fig. 2), 9 million and 230 thousand in early May, and 15 million in late May of 1934 (Fig. 2).   These, combined with the fact that the supply of nitrates and phosphates was considerably lower there in May of that year than in March (Figs. 5, 7), point to April as probably the season of most active production.

The distribution of nutrient salts also suggests a much greater abundance of phytoplankton in April for the eastern part of the Bank (no counts available) than either in March (1 station, 44 million) or in early May (7 stations, average, 5 million; maximum, 6 million), since nitrates declined from 135 mgs. to 4–10 mgs. (Fig. 7).   In this instance too, we find corroborative evidence from the year 1913, when the phytoplankton in the same general region (Lat. 41° 37' N., Long. 67° 18' W.) was so rich on April 26–27 that "the nets soon clogged, although of large mesh" (Bigelow, 1914a, p. 415).

In the southern arm of the western basin in 1934 rich counts were recorded in late April (2 billion) and early May (485 million); whereas in its northern extension the counts for late April (four stations) were slightly smaller even than at one station in March (Fig. 2), and the supply of nitrates and phosphates was about the same (Figs. 5, 7).   The fact, however, that the nitrates were greatly reduced there by late May (Fig. 7) when the phytoplankton was even more sparse than it had been a month earlier, suggests that a

[10] Calculated from the numbers per liter at different levels reported by Gran and Braarud (1935).

peak of abundance had developed and passed by, in the interim between the late April and late May cruises. This is supported by evidence from 1915 when diatoms were still swarming on May 4–14 "over a triangular area in the central part of the gulf extending from Cape Elizabeth to the Grand Manan Channel, and from the coast of Maine at least as far as Cashes Ledge" (Bigelow, 1917, p. 324, Fig. 98). In fact at one station there diatoms were so plentiful on May 4 "that every interstice of the fine net was clogged and its silken bag transformed into a cone of slime almost impervious to water after a few minutes' submergence," the coarse meshed net yielding over 2000 cc. in 20 minutes (Bigelow, 1917, p. 344). In short, it appears by present evidence that the vernal outburst may be expected by the last week in April in the southern part of the western basin, but not until well into May in its northern extension.

It was on the late April 1934 cruise that the largest phytoplankton count was recorded at the level of maximum abundance, i.e., 1,162,000 per liter at 10 meters at a station in the east central part of the basin, where the calculated number per column was about 5,375,000,000. This contrasts with a minimum at the level of greatest abundance (in this case the surface) of only 220 cells per liter in December 1933 in the southwestern part of the basin.

REGIONAL ABUNDANCE OF VEGETATION AT THE CLIMAX OF THE VERNAL OUTBURST

Available data on regional abundance at the climax of the vernal outburst for different localities are:

Locality	Year	Cells per Column
Off Gloucester, Massachusetts	1939	> 2 billion
Coastal belt near Portland, Maine	1934	400–5000 million
Vicinity of Mt. Desert Island	1932	6 billion
Vicinity of Mt. Desert Island	1934	2 billion
Passamoquoddy Bay	1932	4 billion
Mouth of the Bay of Fundy	1932	6 billion
Northeastern part of the gulf generally, including basin as well as coastal shelf	1934	
Average of 9 stations		7 billion
Maximum		14 billion
8 of the stations		> 3 billion
Western arm of basin, southern part		2 billion

The foregoing, contrasted with the mid- or late winter values shown on Fig. 2, indicates that a 1000-fold multiplication of vegetable cells is usual from the winter minimum to the vernal peak in all major subdivisions of the Gulf where the phytoplankton has been sampled near the time when the vernal outburst reaches its climax. The maximum increase for any one locality was, in fact, 60,000 fold, i.e., a few miles east of Portland where the count per column rose from 80,000 cells in December to 5 billion in April (Fig. 2), and for all that is known to the contrary, this may apply equally to the waters of George's Bank where data are lacking at the critical season. As a result of variations in the dates at which the maxima develop and in the rate at which the flora is subsequently impoverished in different parts of .

the Gulf, the regional variations in abundance are wider throughout the spring than at any other time of year.

We must, however, warn the reader that although rich catches are the rule during the season of active flowering, wide variations in the amounts present have frequently been recorded within a few miles; the planktonic flora may even be "so streaky that one can actually see the net pass through alternate bands of brownish diatoms and of clear water" (Bigelow, 1926, p. 403). A similar patchiness was observed off Gloucester, Massachusetts, in early April of 1939, and on the eastern part of George's Bank in April, 1913. There is, of course, nothing novel in the observations that the distribution of the phytoplankton may be so very irregular, for this has often been reported in the shoal water of other seas (Bigelow, 1914a, p. 415). No doubt the fact that very poor catches are occasionally recorded in regions and at times when the counts are generally high, is explicable on this basis. In fact, the distributional chart for late April (Fig. 2) affords one example, i.e., a count of less than 100,000 at one station in the western basin, when all other counts were larger than 1 million.

The range of regional variation in the maximum number of cells that have so far been recorded in the several major subdivisions of the open Gulf at the respective seasons when the vernal outburst of diatoms is near culmination, is only in the ratio of 1 to about 6 or 7; or about 2 billion, contrasted with about 13 billion (Fig. 2). This ratio is perhaps no wider than can be charged to the roughness of our methods, coupled with the chance distribution of the sampling stations, it remaining doubtful whether any station has been sampled when cells were actually the most plentiful there. In the face of this uniformity of results, no one subdivision of the gulf can be classed as significantly more productive than another at the season when the vernal vegetation is most abundant. The Bay of Fundy stands out, however, as a notable exception in this respect, for even though the maximum counts that have been recorded near its mouth are of about the same order of magnitude as for other parts of the Gulf of Maine area, Gran and Braarud (1935) found that the waters of the inner parts of the Bay are far less rich even at the peak season. And it is largely on the basis of this regional contrast that they have based their interesting discussion of factors limiting the production of phytoplankton within the Bay.

The preceding statement that the maximum abundance of phytoplankton is at least of the same general order of magnitude throughout the open Gulf as a whole, should not be taken as implying that the total productivity of the vernal flowerings is equally uniform regionally, for it is necessary to take equally into account the notable contrasts in the duration of the period of abundance of phytoplankton in different localities. In Massachusetts Bay (to recapitulate) this may be only a few days, but several weeks along the coast eastward from Mt. Desert Island, and 1–3 months in Passamoquoddy Bay.

Were we forced to rely on the counts of total phytoplankton alone, we would meet an impasse here, because of uncertainty as to the relative degree to which the persistence of a rich flora in any given locality—Passamoquoddy Bay, for example—may depend on continued multiplication of cells (though at a declining rate) and to what degree on a long period of survival, after the climax of abundance is reached. Neither does Davidson's (1934) analysis of the flora help in this respect, for the same two genera—Thalassiosira and Chaetoceros—that constitute the great majority of the persistant planktonic flora of Passamoquoddy Bay after the vernal climax, are the same two that are chiefly responsible for the development of the latter. Granting these uncertainties it seems, however, permissible to conclude that where diatoms persist at a high and roughly constant level of abundance for as much as 4–12 weeks after the period of rapid multiplication has terminated, as happens in Passamoquoddy Bay, at least a moderate amount of production is taking place; whereas production must be practically at an end after the climax, in regions where the latter is succeeded by as rapid an impoverishment as it is in most other parts of the Gulf.   On this basis, the western coastal belt and the bowl of the Gulf generally would rank relatively low in productivity for the vernal half year as a whole, the northern coastal belt and Passamoquoddy Bay relatively high, also George's Bank; but to attempt any numerical comparison of the relative productivity of these contrasting regions, would be idle from existing data.

LATE SPRING

Preliminary observations for 1915 had already suggested that the vernal peak of production is followed by a decrease hardly less sensational throughout the deep offshore waters of the Gulf as a whole.   This is corroborated by the data for late May of 1934, when four counts only were as large as one million cells per column among 33 stations distributed over the central bowl outside the 100-meter curve, and from southern Nova Scotia out across the Northern Channel and Browns Bank to the Eastern Channel.   Inasmuch as these four were all located on the western side, they no doubt represent the zone of transition to the somewhat richer flora that still persisted in the neighboring coastal belt (Fig. 2). The severity of this impoverishment may be more precisely illustrated by the fact that the late May average for the deep offshore water (27 stations average 834,000) was only about 1/4000 as great as the average for the same general localities for April (3,417,362,000). In fact, the May average for the part of the Gulf in question was only about 1/4 as great as the winter average (3,147,000), a fact interesting because the winter, prior to the vernal enrichment, is usually regarded as the season of greatest planktonic poverty.

Seasonal impoverishment after the vernal climax is also extreme—if perhaps somewhat less so—in most years in the western coastal belt, as exemplified by stations near Cape Ann and Cape Cod.   In 1912, 1915, 1938 and 1939, for example, the vernal peaks—when cells are to be numbered by the billions—were shortly succeeded near Cape Ann by the disappearance of the diatoms and their replacement by a very much sparser peridinian flora.   However, counts of 34 million cells per column close to Cape Cod, and of 2 and 11 million near Cape Ann, as late as the last week of May in 1934 (no data for April) show that at least a moderately rich flora may persist several weeks later in some years than in others on this side of the Gulf.   And although diatoms practically vanished from these waters early in April in 1920, the impoverishment of the flora as a whole which usually accompanies that event was somewhat delayed in that particular year by a great development of Phaeocystis about the middle of the month;[11] however, this flowering persisted only briefly, Phaeocystis having completely vanished by the 4th of May.   Inasmuch as this is our only record for such a flowering in the Gulf, it is probable that such events are sporadic and not a normal stage in the seasonal succession.

Contrasting sharply with the impoverishment, more or less severe, that follows the vernal outburst in the parts of the Gulf just discussed, an abundant flora appears to persist at least locally, until well into the summer in the northeast coastal zone in most years, and

[11] Actual records, April 18–20 (Bigelow, 1926, p. 458).

in waters tributary to the Bay of Fundy. A rich diatom flora was recorded close to Mt. Desert Island, for example, on June 14 of 1915 (no counts were made), and there were 6 billion [12] there on June 28th of 1932, or more than 25 times as many as off Casco Bay some 90 miles to the westward.

Similarly, Davidson (1934, p. 9) found a high level of abundance, though with fluctuations, persisting in Passamoquoddy Bay for several weeks after the vernal climax with "the largest maximum following towards the end of June or in early July," which is in line with Gran and Braarud's (1935) record corresponding to some 600 million cells per column there on June 20th, and of 4 billion on July 30 of 1932. Diatoms also persist longer in abundance in the Bay of Fundy itself than in the open waters of the Gulf to the southward though not as late as in its tributaries, to judge from Gran and Braarud's (1935) counts corresponding to nearly 6 billion per column there on May 23 of 1932. And although they describe an "extreme decline in the phytoplankton" as taking place in the Bay from May to June (Gran and Braarud, 1935, p. 339), the numbers they record at 3 Fundian stations in the latter month (corresponding to 18, 38, and 102 million cells per column, respectively), while much lower than for May, are nevertheless considerably higher than the counts for the open Gulf outside the coastal belt for late May and early June of 1934 (Fig. 2). The available data similarly suggest that a moderate flora, though less rich than the probable April peak, may be expected to persist into early summer in the turbulent waters of George's Bank, to judge from occasional counts of 1–8 million there in early May and of 142 million for late May of 1934 (Fig. 2). Unfortunately we lack numerical data for the extreme regions during the critical season of any one year; therefore it is impossible to calculate the ratios of regional variation, but if the picture for the open Gulf for 1934 is comparable with Gran and Braarud's (1935) counts for 1932, the vicinity of Mt. Desert Island and Passamoquoddy Bay may be expected to support something like two thousand times as rich a flora at the end of May, as does the deep central bowl of the Gulf generally or the waters southward from Nova Scotia.

Summer

The evidence for 1933 and 1934 combined (Fig. 3), is that the planktonic flora of the bowl of the Gulf—at its lowest ebb late in May (average only 834,000 cells per column, p. 161)—may be expected to increase somewhat in numerical abundance through June, counts at 16 stations outside the 200-meter curve having averaged 5,478,000 in the first week of July of 1933, an indication of a 6-fold multiplication in the number of vegetable cells present. This increase, it appears, normally continues through late July, again on the assumption that combination of data for different years be allowable, for Gran (1933) reports counts corresponding to 51–543 million cells per column (average about 272 million) at four stations in the bowl early in August of 1932, or about 50 times as many as were recorded there for early July of 1933 (Fig. 3). Centers of great richness for *Pontosphaera Huxleyi* may also develop locally in the deeper parts of the Gulf in midsummer, witness counts corresponding to 758 million of these tiny plants per column in the southeastern part of the basin on July 16, 1933 (Braarud, 1934).

There is, however, no reason to suppose that this midsummer flowering in the bowl ever approaches in richness the flora that develops there at the time of the vernal peak;

[12] Calculated from an average of 809,946 cells per liter (Gran and Braarud, 1935, p. 452).

the largest August count yet recorded there (543,000,000) is, in fact, only about 1/26 as large as the maximum for April; and the average for the one month only about 1/1630 as great as that for the latter.   And the counts at 7 stations outside the 100-meter contour were significantly smaller (2–41 million, average about 10 million) in September of 1934—the one year of record—in fact only about 1/27 as great on the average as the August count just mentioned.   It appears, in short, that the extreme impoverishment that follows the vernal outbursts of diatoms in the bowl of the Gulf in general is normally succeeded there by a second, but less active flowering of phytoplankton in mid- and late summer.

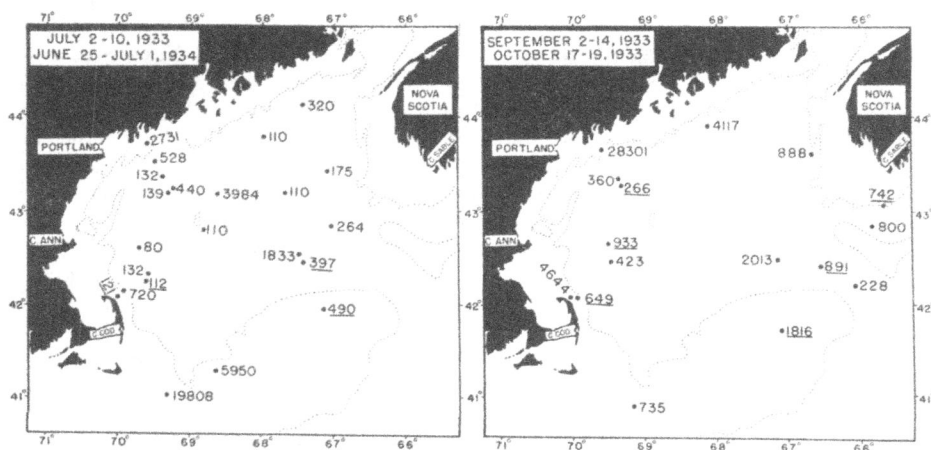

FIG. 3.   Numbers of phytoplankton cells (in tens of thousands) per 0.1 square meter of sea surface. The values for June–July 1934 and for October 1933 are underlined.

It also seems well established that a summer or early-autumn flowering of diatoms takes place generally in the shoaler waters along the coast—one, however, which, by present data, may vary widely from place to place in the dates of its inception, and culmination, as well as in its productivity.   Sceletonema was, for example, recorded in abundance in Massachusetts Bay late in September of 1915 (Bigelow, 1917, p. 326; 1926, p. 404); so abundantly, in fact, that the average catches there had risen to 25–30 cc. per haul, contrasted with only 2–3 cc. at a neighboring locality a month earlier.   And a September maximum is also indicated for the coastal water a few miles east of Portland, by a count there of about 280 million cells per column on the 6th of the month in 1933 (Fig. 3), for which *Guinardia flaccida* was responsible (Lillick, 1940).   Again, a tremendous August flowering of Asterionella (no counts obtained) was encountered near Mt. Desert Island in 1912, extending thence coastwise both eastward and westward for perhaps 30 miles (Bigelow, 1914, p. 133, 1926, p. 431).   In 1932 there were 332,000 vegetable cells per liter at the richest level, and an average corresponding to 359 million cells per column, a few miles off Mt. Desert Island on August 14 (Gran and Braarud, 1935), where there had been only about 1 million per column in the last days of the previous June.   Burkholder's (1933) data for 1930, strongly suggest—if they do not prove—a definite maximum for diatoms in

Frenchman's Bay between Mt. Desert Island and the mainland during late August.   Davidson (1934) has similarly reported an August maximum, more or less well defined, following a decline in July, in 5 out of the 7 years of record for Passamoquoddy Bay, where Gran and Braarud (1935) also had counts corresponding to about 1047 million cells per column on August 16, 1932.   Davidson's (1934) curves for a station (No. 5) just outside the mouth of Passamoquoddy Bay show, it is true, the opposite state, i.e., progressive impoverishment after the passage of the vernal climax in each year of record.   However, the failure of a second flowering to develop just then is seemingly a local phenomenon, for Gran and Braarud's (1935) counts point to an August flora of at least 500–600 million cells per column across the mouth of the Bay of Fundy, for diatoms alone.

Counts of phytoplankton for the coastal shelf west and south of Nova Scotia, for Brown's Bank and for the Eastern Channel are lacking for July and August; nor is any information available as to the supply of nutrients there for those months.   Hence, it is not possible to deduce from the September counts of 2 and 9 million cells (Fig. 3), whether this comparatively sparse flora had or had not been preceded by an August peak of production. High counts have, however, been recorded locally on George's Bank on two occasions in mid- and late summer, corresponding respectively to about 2,814 million on the eastern part on August 2–4, 1932 (Gran, 1933), and to 198 million on the western part in early July 1933 (Fig. 3).   An abundance of diatoms were also encountered, but no counts obtained, on the western end of the Bank on July 9 in 1913, on its northeastern edge on July 23 in 1914, and at 2 stations on the western end on July 23 in 1916 (Bigelow, 1926, p. 391). At least moderately rich production of diatoms is thus to be expected, as earlier surmised, over one part of the Bank or another at any time in July or August, either confined to small areas, as was the case in July of 1914 (Bigelow, 1917), or more widespread, as in the first half of July of 1933 when the flora appears to have been at least moderately rich over the Bank as a whole.

It appears, however, that these summer flowerings on George's Bank are decidedly irregular events, and that they involve the water of the Bank much more generally in some years than in others.   Neither is it "yet clear whether any particular region on the Banks is more favorable for the multiplication of diatoms than another, except that we have always found these rich flowerings on its shoaler parts and never close enough to the continental slope to be within the influence of the high temperature outside the edge" (Bigelow, 1926, p. 391). A rich flora may also exist—with what regularity is not known—in late summer or early autumn over Nantucket Shoals farther west as previously noted, Dr. W. C. Kendall having written "in his field notes that 'on September 2, 1896, the water was very full of brown slimy stuff' at latitude 40° 47′, longitude 69° 43′, which can only have been diatoms" (Bigelow, 1926, p. 391).

In the western coastal belt, where a decided impoverishment succeeds the spring peak, this second period of enrichment resulted in a second and rather prominent peak in the only year when it has come under observation (p. 163).   Along the northern coasts of the Gulf, however, and in the Bay of Fundy this may or may not be the case, depending not only on the productivity of the late summer flowering, but equally on how rich a flora may persist there through late spring and early summer in any particular year, and on what degree of impoverishment may take place prior to the second period of enrichment.   Over the shoal southern rim of the Gulf, by present indication, the second period of flowering is more often made evident by the development of rich centers of circumscribed extent than by a great enrichment involving the waters of George's Bank as a whole.

Available data also suggest significant regional variation in the dates when the second flowering reaches its climax; i.e. that this is to be expected by August in the northern side of the Gulf including the Bay of Fundy, on the one hand,[13] also locally on George's Bank and in the southwestern part of the basin on the other, whereas we have no record of it in the Massachusetts Bay region until a month later.

Autumn

Information as to the numerical abundance of phytoplankton throughout the autumn, subsequent to the second flowering period (the latter falling in late summer or early autumn, depending on the locality), is limited to the counts for September and October shown on figure 3, to the results of towings in the region of Passamoquoddy Bay (Fritz, 1921; Davidson, 1934) and to a few November towings off Gloucester in 1912 (the latter summarized by Bigelow, 1926, p. 395).

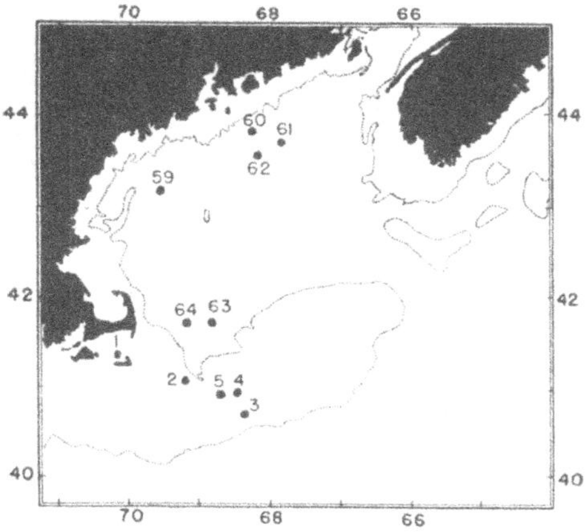

FIG. 4.   Locations of Atlantis stations, August, 1936.

In Passamoquoddy Bay in each of the years when the second flowering developed as early as August (e.g., 1927, 1928, 1931, Davidson, 1934) it was followed either by a rapid depletion leading to a very low level of abundance by the end of the month, or at least by a decline continuing as late into September as the observations were made.   Numerical data for the Bay are lacking for the later autumn, but the history of events elsewhere makes it likely that in years when the second enrichment endures there into September, or throughout that month, e.g., 1917, 1925, 1926 (Bailey, 1917; McMurrich, in Bigelow, 1926; Davidson, 1934), impoverishment follows soon thereafter.   The chief annual differences in this respect for the Bay are in the date at which impoverishment commences, and in its severity.   Average counts corresponding to about 95 million cells (6 stations, maximum,

[13] Horizontal tows in August, 1936 yielded an abundance of diatoms at the stations marked on Fig. 4.

363 million) for the open Bay of Fundy for September 10–15 of 1932 (Gran and Braarud, 1935, Tables), contrasted with the comparatively high counts of the previous month (p. 164), again show impoverishment sufficient to have reduced the flora to about 1/5 or 1/6 of the abundance that prevailed there shortly previous.

A similar and coincident depletion is equally indicated for the bowl of the Gulf (outside the 100-meter contour) as a whole—witness counts there equivalent to 51–543 million per column (average, 272,000,000, 4 stations) in August of 1932, but of only 2–41 million (7 stations, average, 10,026,000) in September of 1934, and of 2–9 million in October of that year (Fig. 3). A rather sparse flora (7–18 million) was also recorded then on George's Bank, contrasting with the frequent rich concentrations of mid- and late summer (see above); and although the second flowering may not develop in Massachusetts Bay or in the coastal belt as far north as the offing of Portland until well into September (p. 163), the phytoplankton in the vicinity of Gloucester proved, in 1912, to be very scanty indeed by the last week of November (Bigelow, 1914a, p. 405; 1926, p. 395).

Data are lacking for November for other parts of the Gulf, but the very low counts recorded for December, 1933 (p. 155, Fig. 2) make it extremely unlikely that rich concentrations of phytoplankton of any sort normally develop anywhere in the Gulf in the last two months of autumn. This is corroborated by the fact that in 1920 also "the several species of diatoms that are most abundant from spring to early autumn had practically vanished from the whole coastal belt between Cape Cod and the mouth of the Bay of Fundy by December" (Bigelow, 1926, p. 395)—to be succeeded by Coscinodiscus which has never been found in more than moderate numbers in the Gulf (Lillick, 1940).

NOTES ON SOME FACTORS AFFECTING THE ABUNDANCE OF PHYTOPLANKTON IN THE GULF

It is obvious that the significant fact in regard to abundance is not so much the number of vegetable cells present from time to time—for among different planktonic plants these vary widely in size—but the mass of living matter produced. Even as rough a comparison between numbers of cells and total volumes as available data have so far afforded for the Gulf of Maine has been enough to show that the two are not parallel, though the trend from season to season is similar. It has been suggested that a somewhat better picture may be afforded by measurement of the amount of chlorophyll present and of carotin pigments, per unit volume of water, taken as an index to the production of carbohydrates,[14] a method not yet applied to the Gulf of Maine. Granting all the shortcomings of counts of planktonic plant cells as a basis for studies of production, they do give at least a rough picture of the ups and downs of the planktonic flora from season to season, the range of numerical abundance being so wide as to make it unlikely that the total mass of phytoplankton is ever large, when the number of cells is small, or vice versa. Comparison between the numbers of cells and the amounts of phosphates and nitrates in solution in the water—or any other factor that may be chosen for analysis—seems therefore allowable as a rough index of the degree to which total production of the phytoplankton is correlated with the latter.

Gran and Braarud (1935, p. 404) have already shown, in their investigation of 1932, that "the general changes in the growth and decline of the phytoplankton population of the

[14] For examples of the application of this method and for discussions of its validity, see especially Harvey (1934), Riley (1939, 1939a), Krey (1939), and Gillam, El Ridi, and Wimpenny (1939).

Gulf of Maine . . . agree with the generally adopted theories, that the quantitative varia-tions are governed by the supplies of light and nutrient salts (phosphates and nitrates)" with light, of course, determining the thickness of the productive stratum of water. The data for 1933–1934 not only corroborate this generalization, as was to be expected, but allow a somewhat more precise evaluation of the effects of certain of the factors involved.

INITIATION OF THE VERNAL OUTBURST

The outstanding points to be considered in relation to the planktonic cycle of the Gulf are: first, the stimulus for the vernal outburst of diatoms, and for its occurrence at a par-ticular date; second, the cause of its eclipse after a longer or shorter period of active produc-tion; third, the causes for the regional variations that exist in the numerical abundance of the planktonic flora, diatoms and peridinians combined, during the summer and early autumn; and fourth, the causes for autumnal impoverishment of the flora. The factors involved include (1) the chemical, dissolved nutrients,[15] (2) the physical, i.e., light, temper-ature, activity of vertical circulation, and (3) the degree to which the abundance of planktonic plants is affected by the animals that graze upon them.

The general rule that the diatoms, at least, depend for active flowering on an adequate supply of nutrients in the water is now generally accepted. Equally, however, it has been established that in some parts of the sea active flowerings may not develop until several months after the supply of nutrients has built up to a high level in the depth-zone of plant production. We may state at once that comparison of the cell counts for the year 1933–1934 with the coincident determinations of phosphates and nitrates proves that the Gulf of Maine as a whole falls in that category. To support this generalization it is perhaps enough to point out that the concentrations of nutrient salts were maximal (ca. 138 mgs. of nitrate and 110 mgs. of phosphate (PO_4) per cubic meter, Figs. 5, 7) in the upper strata of the Gulf as a whole in December and January; whereas active flowerings have never been recorded earlier than the beginning of March except in very restricted localities (p. 155). In the occurrence of a lag of this sort and one of such considerable duration, the Gulf agrees with certain Norwegian localities (Føyn, 1929; Gran, 1930; Braarud and Klem, 1928) and with Puget Sound (Gran and Thompson, 1930).

To explain such a situation it has been suggested that vernal flowerings await the in-crement of nutrients contributed by an inrush of river water (Nathansohn, 1910; Gran, 1933; Bigelow, 1926; Ercegović, 1940). In the Gulf of Maine, however, the entry of fresh water can hardly have been a controlling factor in 1934, for salinities did not reflect any great vernal freshening until May, i.e., not until long after the onset of the spring diatom flowering. It seems still less likely that the late summer flowerings along the north and northwest coasts of the Gulf owe their existence to the nutrients contributed by the drainage of that region, river-discharge being very small in volume at that time of year. Further-more, rich flowerings occur on George's Bank, far from any immediate influence from this source.

These generalities argue in particular against the view that the vernal outbursts in the Gulf await the soluble iron from the land that may be contributed by the river-freshets of spring. Having no determinations to contribute of the amounts of iron in solution, we need only refer in this connection to Gran's (1933) finding from culture experiments at

[15] For recent discussion of growth requirements of diatoms, see Harvey, 1939.

Woods Hole, that while the addition to the culture of soil extract and of a ferri-ligno-protein did stimulate the reproduction of certain neritic diatoms (*Leptocylindricus danicus*, *Nitzschia seriata*), it did not significantly affect that of the oceanic species, *Rhizosolenia alata*, and the diatom species responsible for the mass flowerings of the Gulf of Maine chiefly belong to the "planktonic" category.   Neither do such measurements as have been made of the amounts of silicates in the Gulf of Maine (Bigelow, 1926) or elsewhere, suggest that these are ever apt to be a stimulating or a limiting factor in the Gulf.

In short, our data lead to the general conclusion that though an augmenting flora cannot occur without a sufficiency of nitrates and of phosphates, it usually is not the attainment by these two nutrients, nor of any other [16] of any particular richness in the surface waters that stimulates the vernal outburst of plant cells in the Gulf of Maine.

The sporadic occurrence of rich flowerings of diatoms in Cape Cod Bay in December (p. 155), added to the fact that at Woods Hole the chief diatom maximum comes in that same month in most years (Fish, 1925), is good evidence that in the Gulf—unlike the English Channel (Harvey, 1926; cf. also Marshall and Orr, 1926)—illumination is strong enough for active photosynthesis, even in early winter; as strong in fact, when it is at its minimum for the year, as it is in Norwegian waters in spring, i.e., at the time of the vernal outburst (Kimball, 1928).   And the possibility that differential light-requirements for different diatoms enter into this particular case is barred, since the same species are dominant in the Gulf, as off Norway.   The fact that there is as much as six weeks variation in the dates of the vernal outburst in different parts of the open Gulf—a difference far wider than could be credited to differences in total illumination resulting from the differences in latitude between the stations in question, or from local differences in cloudiness or in the prevalence of fog—is further evidence that the factor delaying the rapid multiplication of diatoms is not any insufficiency of light, except perhaps in the Bay of Fundy, where local conditions are peculiar owing to the high degree of turbidity of the water, as shown by Gran and Braarud (1935).

Neither does it appear that vernal flowerings are conditioned by any particular value of temperature, for the surface waters are somewhat warmer over the Gulf as a whole in December when the phytoplankton is very sparse indeed, than in April, when diatom flowerings are at their height over most parts of the Gulf.   It is, of course, possible that a rising temperature has some stimulating effect in the Gulf where the surface cools generally to its yearly minimum about the last of February or the first of March.   But without laboring the point we may say at once that we have not been able to find any definite correlation—positive or negative—between regional variations in the seasonal schedule of the flowerings of the planktonic flora as a whole or their productivity, and the regional variations in the temperature of the surface waters, although such a correlation undoubtedly exists between seasonal variations in the temperature and the succession of dominant species.

Particularly interesting in this connection, is the fact that in 1934 diatoms continued at a low level of abundance throughout April in the western basin where the concentrations of nutrients had been highest in March; whereas in the eastern basin under essentially the same conditions of temperature and light, April flowering was intense.   Nor can the longer

[16] Concentrations of nitrites and ammonia were so low throughout the year 1933–1934 that no attempt has been made to relate phytoplankton to them.   For an account of these the reader is referred to Rakestraw (1936) and Redfield and Keys (1938).

lag in diatom production in the first of these localities be laid to the score of insufficient "seeding," since the species responsible for the vernal flowerings elsewhere, were also present there at the time.

Summation of available evidence thus leads to the conclusion that in the Gulf, as in sundry other localities (Gran, 1932, p. 351; Gran and Braarud, 1935), something more than a mere adequacy of nutrients—and something less obvious than any particular temperature or degree of illumination—is needed before the vernal outburst of diatoms can take place.

Vertical Stability of the Water Column. Gran and Braarud (1935) have already shown that although active turbulence of the water favors the production of planktonic plants in shoal regions such as George's Bank, it may also prevent the development of rich flowerings if the turbulent layer through which the vegetable product of the illuminated zone is dispersed downward, be of any considerable thickness. It is on this basis, and with a high degree of probability, that they explain the rather poor production of phytoplankton in the Bay of Fundy, in the face of the rich supply of nutrients that prevails there throughout the spring and summer. The data for 1933–1934 now afford an opportunity to test the efficacy of turbulence prior to the vernal stabilization of the water, as a limiting factor for the open Gulf.

The most hopeful line of attack on the problem is to follow the coincident changes in floral abundance and in the degree of stabilization, at the localities where records were obtained both before, during and after the vernal outburst. In the table herewith, stability is stated as the difference in density [17] between the surface and the 30 meter level.

The localities, representative of the major subdivisions of the Gulf, fall into several categories. The simplest situation is that existing on George's Bank where, corroborating Gran's (1933) and Braarud's (1934) summer data, flowerings of at least moderate richness (up to 142,000,000 per column) can develop in spring, when the water column has no measurable stability, but where the total depth of water (20–80 meters) does not greatly exceed that of the zone of production. A rich flowering was similarly recorded in 1925 in the shoal waters of Cape Cod Bay (maximum depth, 53 meters) in December when the water had either no stability, or even a negative stability.

The situation is equally clear-cut, but antithetical, at the sampling location off Mt. Desert Island where the vernal outburst did not occur until the upper 30 meters had become stabilized, following a period (no data for January) during which the waters had a low stability. And the situation in regard to nutrients suggests that this order of events applies equally to the southern part of the western basin, where the cruises missed the vernal flowering but where the very low values of 17 mgs. of phosphate and 4 mgs. of nitrate early in May (Figs. 5, 7) fix this as having occurred shortly previous, i.e. when the difference in density between the surface and 30 meters was at least 0.2; whereas in March, January and December it had been only 0.01, 0.04 and 0.01, respectively. In other cases, however, in which the sequence is basically similar the preflowering period of instability is less protracted, being confined either to March, as in the coastwise area near Seguin Island and in the southeast part of the basin, or to January, as in the eastern side of the Gulf near Lurcher Shoal and in deep water in the northwestern part, to judge from the situation as regards nitrates and phosphates.

[17] Density = 1000 (specific gravity − 1).

Locality	Month	Stability	Phytoplankton Cells per Column
Eastern George's Bank	Dec.	—	4080
	Jan.	0.02	2618
	March	0.02	44,110,000
	Early May	0.03	9,000,000 (av.)
	Late May	0.00	1,520,000 and
			142,000,000
Western George's Bank	Dec.	—	—
	Jan.	0.00	18,084,000
	March	0.00	27,860,000
	April	—	—
	Late May	0.01	15,360,000
South Eastern Basin	Jan.	0.22	1,859,000
	Mar.	0.01	550,000
	Early May	0.09	2,121,850,000
North Channel	March	0.10	26,730,000
	Early May	0.01	489,850,000
Western Basin, North	Dec.	0.12	1,620,000
	Jan.	0.04	4,675,000
	March	0.02	3,245,000
	April	0.17	1,260,000
	Late May	—	730,000
Western Basin, South	Dec.	0.01?	570,000
	Jan.	0.04	3,498,000
	Mar.	0.01	853,000
	April	0.23	80,000
	Late May	—	9,000,000
Near Cape Cod	Dec.	—	3,120,000
	Jan.	0.05	7,216,000
	Mar.	0.15	14,493,000
	Early May	0.30	8,840
Lurcher Shoal	Dec.	0.10	3,230,000
	Jan.	0.04	1,672,000
	March	0.15	25,630,000
	April	0.19	2,750,750,000
Off Portland	March	0.03	60,655,000
	April	2.91	447,090,000
Off Seguin Island	Dec.	0.42	80,000
	Jan.	0.59	5,269,000
	Mar.	0.13	117,425,000
	Apr.	0.90	5,056,550,000
Near Mount Desert Island	Dec.	0.36	550,000
	Mar.	0.03	2,805,000
	Apr.	0.22	3,238,850,000

Thus it appears that rich flowerings did not develop in certain sections of the Gulf, mainly in the bowl, in the spring in question until a positive stability of approximately 0.1 had built up in the upper 30 meters.   Lack of stability of the water column with consequent turbulence may then perhaps be classed as one of the factors which, in conjunction, are responsible for the very considerable lag between the date when the water is fully restocked with nutrients and that when the vernal outburst of vegetation occurs.   However, it is not possible to explain on this basis the situation existing in the northern coastal zone as illustrated by the station near Seguin Island (see preceding table) where the water was decidedly stable throughout the period from December to April, but where the lag was as long and the contrast between autumn-winter poverty and vernal richness of vegetation as sharp as it is anywhere in the Gulf.   Neither does the state of stability of the water offer any explanation for the failure of productive flowerings to develop in the coastal belt in the Nova Scotian side of the Gulf (e.g., near Lurcher Shoal), in the deep water in the northwestern part, on the eastern part of George's Bank in December, or in the southeastern part of the basin in January, for the water had about the same stability at those localities (or lack of it in the case of George's Bank) in those months and with an abundance of nutrients as it had at the time of the spring flowerings.

We can only conclude then, that while under certain conditions a lack of stability may operate to prevent active flowerings in the deep parts of the Gulf generally during the winter months, this is only part of the story, and that some other coincident factor acts in the same direction, though clearly not with universal efficiency—witness the December flowering recorded above (p. 155) for Cape Cod Bay.   This factor, we suspect, is to be sought in the effects on the flora of grazing by herbivorous members of the animal plankton, the importance of which has been stressed of late, especially by Harvey (1934a).[18]

EFFECTS OF GRAZING

Unfortunately we are in no better position than were Gran and Braarud (1935, p. 422) to estimate the effect of grazing.   Earlier surveys (Bigelow, 1914a, 1926) point, however, to an autumnal increase in zooplankton in the Gulf so pronounced that vertical hauls yielded on the average more than twice as much in the Massachusetts Bay region in October of 1915 as in the preceding August (Bigelow, 1926, p. 87).   The planktonic population of animals then decreases through late autumn and early winter, to reach its minimum over the Gulf generally, by the last part of February or first days of March.   The Gulf of Maine thus falls in the category of localities where the vernal maximum for planktonic plants shortly succeeds the yearly minimum for planktonic animals—a relationship the significance of which is emphasized by Fleming (1939); and the evidence of a causal relationship between the mutual fluctuations of the two groups is as strong as could be expected without precise numerical information as to the abundance of herbivores and their consumption rate.

TERMINATION OF THE VERNAL OUTBURST

Phosphates and Nitrates. Gran and Braarud (1935) have concluded that the chief factor limiting the production of phytoplankton in the open Gulf during the summer, i.e. after the climax of vernal production, is the low content in nutrient substances of the surface

[18] For a recent discussion of this subject, with calculations of its importance as a determinant of plant abundance, see Fleming, 1939.

layers that results from the draft made on them by the planktonic plants, coupled with the hindrance to their renewal from below that results from the aestival stratification of the water column.   They have further pointed out that "where the stratification is broken as on George's Bank, and near the coast a richer production is possible also during the summer"

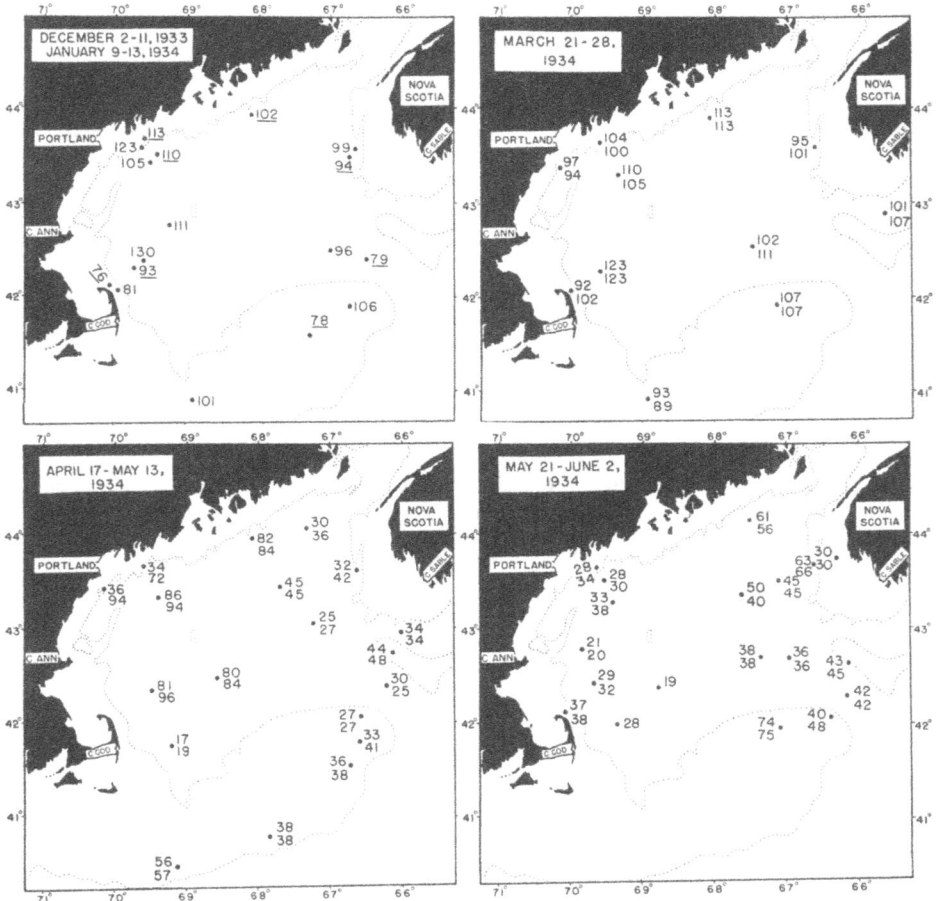

FIG. 5.   Milligrams of PO₄ per cubic meter, at 10 meters for December (underlined) and January; and at the surface (above) and ten meters (below), or surface alone, for the other months.

(Gran and Braarud, 1935, p. 403).   They also concluded that in the particular year (1932) when their studies were carried out, phosphates were more important than nitrates as a limiting factor.   In 1933–1934, however, the relative importance of these two classes of compounds was clearly the reverse in influencing vegetable production.

The vernal outburst of diatoms results, it is true, in a very considerable reduction of phosphates in the upper water strata (Redfield, Smith and Ketchum, 1937) as appears from a comparison of the values for December, January and March with those for April and May (Fig. 5). This indeed was to be expected. Neither is there anything astonishing in the fact that it was in the eastern side of the Gulf where the production of plant cells was the most active from March through April (p. 157) that the coincident reduction in phosphates was the most severe, i.e. from an average of about 107 mgs. of phosphates (PO_4) at 10 meters in March to 42 mgs. per cubic meter in April. Gran and Braarud (1935, p. 395) have, in fact, shown that a much more severe impoverishment may ensue locally, having recorded two instances of total exhaustion of phosphates in the upper 10 meters off Casco Bay and off Mt. Desert Island in May, with coincident values as low as 3–14 mgs. (PO_4) [19] per cubic meter at neighboring stations; 4–15 mgs. in the same general region in June; and 9–45 mgs. in August of 1932.

On the other hand, the late May chart for 1934 (Fig. 5) shows that in that particular year very considerable quantities of phosphates (average 40 mgs.; maximum, 75 mgs.) still persisted in the upper stratum after the eclipse of the major outburst of vernal vegetation with no evidence that even the most active centers of production had resulted in a reduction of phosphates below 20 mgs. per cubic meter at 10 meters, or below 21 mgs. at the

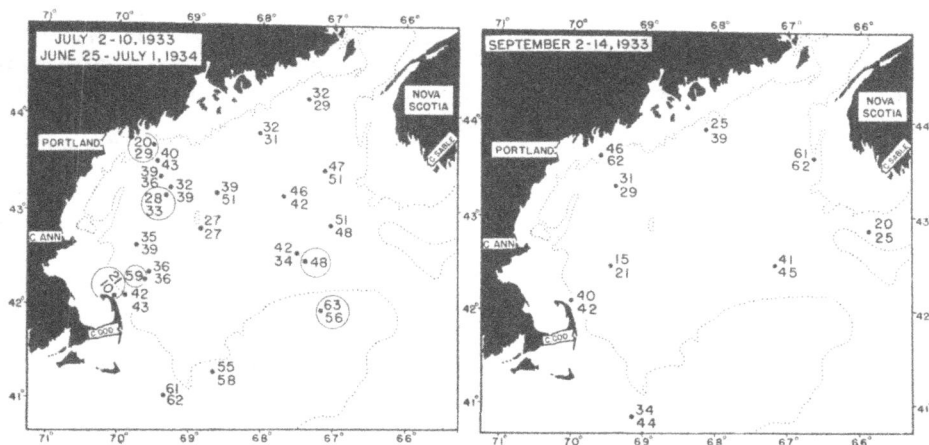

FIG. 6.   Milligrams of PO_4 per cubic meter at the surface (above) and 10 meters (below).
Values for June–July 1934 encircled.

surface over any considerable subdivision of the Gulf. The charts for the summer and autumn (Fig. 6) reveal one case where the phosphate value fell as low as 20 mgs. at the surface, averages at 10 meters being 29 mgs. for June, 39 mgs. for July, and 36 mgs. for September, even if the richer supply on George's Bank be left out of account. And we may say at once that in 1934 the concentration of nitrates during the season when it was scarcest

[19] For comparative purposes, all phosphate values published by other authors are stated here as PO_4.

was very seldom as much as 10 mgs. less at the surface than at 10 meters, which is well above the compensation depth for phytoplankton during the vernal-aestival half year (Gran and Braarud, 1935, p. 403).   Rakestraw (1933) also found 13–34 mgs. at 10 meters at 8 out of 9 stations in the basin of the Gulf and on George's Bank in early August of 1932, with reduction to 8 mgs. at one station only.

Data for the two years combined thus indicate that although the growth of phytoplankton may result in extreme impoverishment of phosphates locally and temporarily, both in spring and in summer, it is more usual for phosphate values to continue higher than about 20–30 mgs., over the Gulf generally, throughout the vernal half year even at the stations where the greatest production of planktonic plants takes place.   We are thus faced with the question whether or not a concentration of phosphate as great as about 20–30 mgs. per cubic meter is likely to be less than sufficient to support a (numerically) strong population of planktonic plants of one species or another: in other words, of what importance as a causative factor is a reduction in phosphate of that amount likely to be in relation to the very severe impoverishment in planktonic vegetation that takes place over so much of the Gulf between April and the last of May (p. 161).   To attempt a categorical answer would of course be premature, pending some definite knowledge of the nutritive requirements of the particular genera of plants concerned, mainly Thalassiosira and Chaetoceros.[20] It is, however, suggestive that the zone of production was nearly as rich in PO_4 in the Gulf, at its poorest, as it is at its richest in the English Channel, where reported values range between 40 and 75 mgs. per cubic meter just prior to the vernal outburst of vegetation (Atkins, 1925–1929; Harvey, 1928a; Cooper, 1933, 1933a, 1938; Harvey, Cooper, Lebour, and Russell, 1935).

Near Møre, also, on the west coast of Norway the highest phosphate value reported at 10 meters by Braarud and Klem (1931) was 67 mgs., the average 42 mgs., while the maximum at 25 meters was only 69 mgs. at any time of year.   Similarly, Marshall and Orr (1926) had between 53 and 67 mgs. only, in March at 10 meters, and about 53 mgs. in February at one locality, with about 53 mgs. at a second, in the Clyde Sea area (west coast of Scotland), preceding the April outburst of diatoms, with July values at 10 meters (preceding the August flowering) of 40–47 mgs. and 53–60 mgs. respectively, at these two localities.

Phosphate concentrations of 40–75 mgs. PO_4 per cubic meter being thus proved sufficient to support intense flowerings of diatoms in seas comparable in depth, temperature and salinity, there is no apparent reason, so far as phosphates are concerned, why the bowl of the Gulf of Maine should not have supported a much richer flora from May onward through the summer of 1934, than was actually the case.   This is in line with Gran and Braarud's (1935) conclusion that phosphates are seldom—if ever—a limiting factor in the inner part of the Bay of Fundy, where most of the values for the 10 meter level for May, June and August of 1932 fell between 13 and 50 mgs.   This, according to Davidson (1934, p. 43), applies equally to the Passamoquoddy region where the minimal phosphate values recorded for the period August 1930–October 1931 were 36 mgs. within the Bay, and 24 mgs. off its mouth.

In short, it is doubtful whether the growth of phytoplankton was seriously limited over any extensive subdivisions of the Gulf at any season by any widespread scarcity of phos-

[20] For a discussion of the nutritive requirements of *Nitzschia Closterium*, see Ketchum, 1939, 1939a.

phates in 1933–1934. But we have no basis for judging whether the low values recorded locally in the open Gulf in 1932 by Gran (1933) and by Gran and Braarud (1935), contrasted with those for 1934, represented an annual difference involving the Gulf as a whole, or whether centers of equally severe impoverishment for phosphates may have developed here and there in 1934 also, but were missed by the grid of stations.   This latter supposition seems, however, the more likely of the two, i.e., that even though phosphates be usually in sufficient amount over the Gulf as a whole, they may be so exhausted by flowerings of great intensity in particular localities where vertical circulation is weak, as to become a severely limiting factor locally, and for a time.

The vernal outbursts of 1934 reduced the average concentration of nitrates at the 10 meter level from 133 mgs. per cubic meter in March to 44 mgs. in late April–early May, to 18 mgs. by late May to early June (Fig. 7), and to 8 mgs by the last days of the latter month, when two cases of total exhaustion were recorded for the 10 meter level (Fig. 7), and three for the surface.   Although the more comprehensive July survey for 1933 (Fig. 8) failed to reveal any values lower than 9 mgs. at 10 meters or lower than 8 mgs. at the surface, the general mean at 10 meters for that month was only 14 mgs., and the maximum only 20 mgs.   The combined record thus makes it clear that in the open Gulf of Maine pelagic plants made a much more severe draft on the supply of nitrates within the photosynthetic zone during their vernal multiplication than they did on the phosphates—a draft that finally resulted in total exhaustion of the salt in question over certain areas, especially in the basin of the Gulf—something that has rarely been recorded for phosphates. In this respect, as Gran and Braarud (1935) have already emphasized, the Bay of Fundy contrasts strongly with the open Gulf, for while plant growth considerably reduces the richness of the surface stratum in nitrates in the Bay as well, below the probable February–March value of more than 100 mgs., the impoverishment is not as extreme as in the open gulf, the minimum value recorded in their tables being 19 mgs. even at the surface, with average surface values of 43 mgs. for May–June (5 stations); 41 mgs. for August, and 31 mgs. for September (3 stations).   At the same time considerable evidence has accumulated to the effect that rich flowerings of planktonic plants demand on the whole a richer supply of nitrates than of phosphates, for we have found no case recorded where an active vernal outburst of diatoms is not preceded by a nitrate concentration of at least 40–50 mgs. per cubic meter at some level within the photosynthetic zone.   In the English Channel, for example, Harvey (1928a) found 60–80 mgs. at the surface in February and March.   Braarud and Klem (1931) had upwards of 40 mgs. in March of 1929 and 1930 in 12 out of 14 determinations at 10 meters and at the surface (average more than 63 mgs., maximum more than 128 mgs.), in Norwegian waters near Møre, with 30 and 53 mgs., respectively, at 25 meters in two other cases, when the 10 meter values were low (14 mgs. and 6 mgs.). Gessner (1933) reports surface values averaging 92 mgs. at 11 stations along a profile crossing the Baltic on February 19, of 1933, i.e., before diatoms began to propagate actively. And Moberg's (1928) graph showing somewhat more than 50 mgs. per cubic meter off La Jolla, California at the 30 meter level where diatoms were the most abundant, but with the surface deficient in both nitrates and diatoms in measurable amount, is of similar import.

In most other localities where the nitrates have been measured at a date shortly preceding rich outbursts of vegetation, the concentrations found have been even higher; in the Barents Sea, for example (Kreps and Verjbinska, 1930); in Puget Sound (Phifer and Thompson, 1937); and proverbially so in the Antarctic (Ruud, 1930). On the other hand, what few determinations have been made, in parts of the ocean where there are no characteristic peaks of abundance for phytoplankton, have shown very low values for nitrate in the illuminated zone. Cases in point, using modern methods, for the open oceans are found in

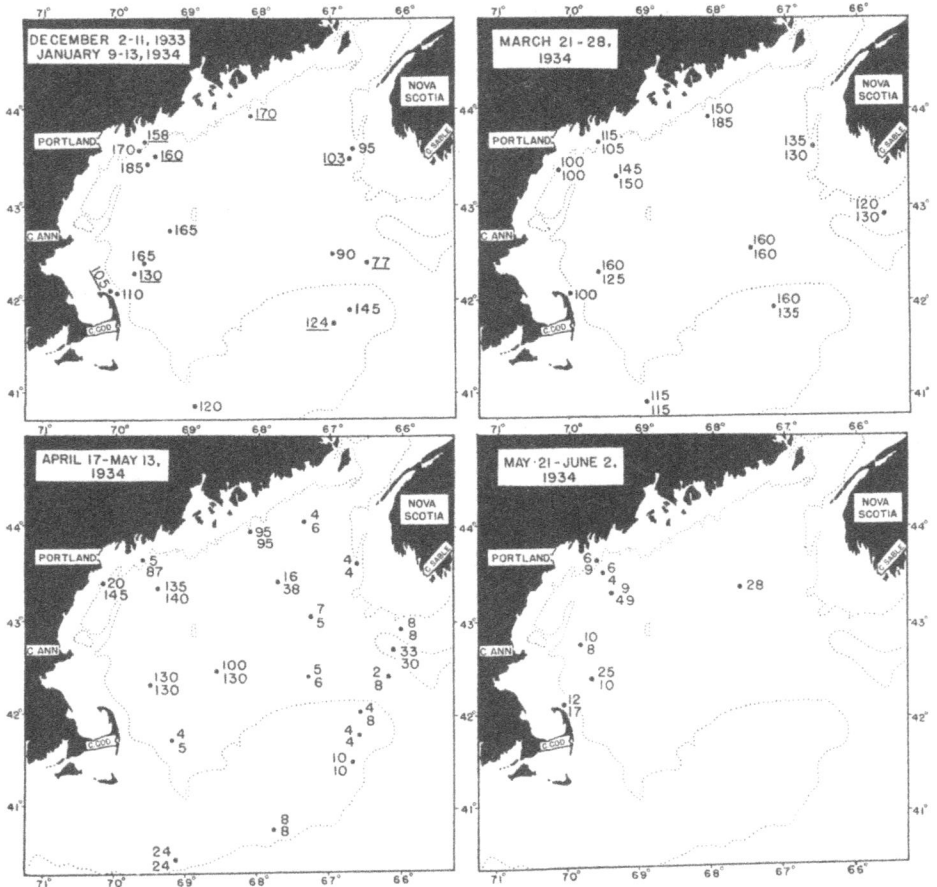

FIG. 7. Milligrams of nitrate per cubic meter at 10 meters for December (underlined) and January; and at the surface (above) and 10 meters (below), or surface alone, for the other months.

the South Atlantic (20 mgs., Ruud, 1930), off the coast of Portugal (11–15 mgs., Harvey, 1928a), and in the Caribbean and Sargasso Seas (Rakestraw and Smith, 1937).

A reduction of the nitrate concentration below 20–30 mgs. per cubic meter in the illuminated zone such as follows the vernal outburst of diatoms over the Gulf of Maine generally, might therefore be expected to act as a definitely limiting factor for plant growth

—at least for diatoms—and still more so the total exhaustion that involved such considerable areas in the early summer of 1934 (Fig. 7).

Without attempting to proceed more deeply into the matter (available information is not sufficiently detailed) the evidence is strong that impoverishment of nitrates, which contrast in this with phosphates, by consumption during the vernal outburst of diatoms,

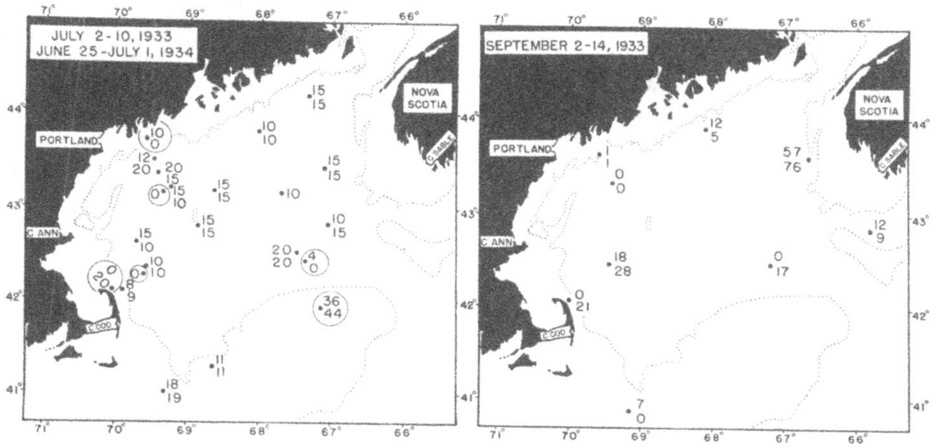

FIG. 8. Milligrams of nitrate per cubic meter at the surface (above) and at 10 meters (below). Values for June–July 1934 encircled.

was the chief cause for the subsequent decline in abundance of planktonic plants over the Gulf as a whole in the particular year in question. We incline also to the belief that while there may be a considerable variation from year to year in this respect, the reduction of nitrates is more likely to be the limiting factor of the two on the whole in the Gulf, during the season when renewal from greater depths is severely limited by the strongly stratified state of the water, because the nitrates from decomposing organic matter are returned more slowly to the water than are the phosphates, there being a greater number of intermediate steps in the process (Cooper, 1937, 1937a).

In the inner parts of the Bay of Fundy on the other hand Gran and Braarud (1935, p. 430) have shown that the limiting factor is not the supply of nutrients of any sort since nitrates like phosphates are in sufficiency there right through the season, but the extraordinary turbidity which limits the penetration of light, thereby reducing the thickness of the zone of production, combined with great turbulence that rapidly disperses the vegetable products of that zone downward through the entire water column (p. 169).

The effects of grazing must as certainly be taken into account, in connection with the impoverishment of the flora, following the vernal climax (when such occurs) as it is in connection with the situation existing in winter. Again, we can make no estimate of this effect beyond noting that in April, when the phytoplankton was locally so abundant that many of the amoeboid and ciliate protozoa were found literally packed with the cells of Thalassiosira.

REGIONAL VARIATIONS IN PLANT PRODUCTION IN SUMMER

It has long been appreciated that in regions where the transparency of the water is normal, active turbulence tends to favor the production of phytoplankton by constantly bringing up fresh supplies of nutrients from the deeper layers. And it is on this basis that Gran and Braarud (1935) have explained the fact that a rich production of diatoms continues into the summer in the coastal belt eastward from Penobscot Bay as well as on George's Bank. Actually, however, the concentration of nitrate recorded by them in August of 1932 was somewhat lower at the surface close to Mt. Desert Island (32 mgs.) than it was over the basin to the southwest (43 and 48 mgs.) where the flora was much less rich (Gran and Braarud, 1935, Fig. 40). Neither is the evidence clear in this respect for George's Bank, since the stock of nitrates was about the same in the basin of the Gulf, where the phytoplankton was scarce, as on the eastern part of George's Bank, where it was fairly plentiful at the time of the August 1932 and July 1933 cruises; while in June–July of 1934, when the phytoplankton was about as abundant on the northeast part of George's Bank as it was in the neighboring parts of the basin to the north, nitrates were considerably richer in the one locality than in the other (Fig. 8).

The data are not sufficiently detailed as to date or as to location to show how generally the expected correlations between plant abundance and the supply of nutrients applies at times and places where flowerings of diatoms take place late in the summer or early in the autumn. However, Rakestraw's (1933) report of 20–27 mgs. of nitrate at 10 meters in the basin, and on George's Bank as well, early in August of 1932, with Gran and Braarud's (1935) of 32–48 mgs. at 10 meters and at the surface on the northern side of the gulf in that same month, with one value as high as 76 mgs. at 10 meters in September of 1934 (Fig. 8) suggests that in mid- and late summer there is a prevailing tendency for nitrates to be supplied to the photosynthetic zone from below faster than plants consume them.

It is also suggestive in this connection that the concentration of nitrate was considerably higher at a station near Lurcher Shoal on the Nova Scotian side of the Gulf at the time of the September cruise of 1933 than anywhere in the basin to the westward or on George's Bank (Fig. 8), but with no significant difference to correspond in the number of plant cells. The obvious inference is that the late summer diatom maximum which develops along the northern coastal belt and over the southern banks of the Gulf as well, affects the nutrient supply not only in these particular regions but over the bowl as a whole. On the contrary the coincident presence of such a considerable quantity of nitrates in the surface waters of the eastern side of the Gulf is a reason for doubting whether the second flowering extended to the coastal waters off Nova Scotia at all in the year in question; whether, in fact, it is a regular item in the seasonal cycle there (p. 164).

AUTUMNAL IMPOVERISHMENT

The records for September 1933 show that the second flowering of late summer or early autumn may again reduce the nitrates to the vanishing point here and there; witness the two instances of total exhaustion indicated on Fig. 8, and a third where 1 mg. only was recorded at the 10 meter level, with only 5 and 9 mgs. at two other stations. Phosphates also had been reduced to 21 mgs. at one station in that month, to 25 mgs. at another. Without data for October we have no way of knowing how soon this tendency toward depletion becomes more than nullified by an increasing supply from below, consequent on the decreasing stability of the water column—certainly this happens before the end of the autumn as proved by the high concentrations of nutrients at the 10 meter level in December (Figs. 5, 7). However, the instances just stated are enough to show that nitrates (perhaps phosphates) may as sharply limit the plant outbursts of early autumn as they do those of the spring (p. 178). At the same time, it is probable that the draft made upon the flora by the planktonic animals that feed upon it increases in severity during mid-autumn, for available data suggest that a diminution in the volume of zooplankton during the summer is "followed by an autumnal increase, which was so considerable in 1915 that there was over twice as much plankton per square meter in water only 80 meters deep by the end of October," in the Massachusetts Bay region as at a neighboring locality in 140 meters in August, with evidence that this augmentation was general throughout the western coastal belt at least (Bigelow, 1926, p. 87, 88). Increasing activity of vertical turbulence may also act adversely in some localities as early as October, for the difference in density between the surface and the 30 meter level was only about 0.07 in October of 1933 in the northwestern part of the basin, where it had been about 2.37 in the previous month and about 1.62 in July, but it is not likely that the break-down of stratification becomes generally effective as a limiting factor until mid-winter (for further discussion of this factor, see p. 169).

Summary

The phytoplankton of the Gulf is scanty (the familiar winter minimum) from late autumn until early spring. In the one year of record the minimum was reached in December in the western coastal belt, and thence southward across the southern rim of the Gulf, but not until January in its eastern side generally. The counts for 1933–1934 point to some 80,000 cells per column as about the minimum, some 18 million per column about the maximum, and about 4 million per column about the average to be expected for December and January combined in a normal winter (Fig. 2). Local flowerings have however been encountered in the Massachusetts Bay region in December (p. 155), and on the western part of George's Bank in January.

In most years the numbers of vegetable cells (diatoms chiefly responsible) increase considerably in the western coastal belt and on western George's Bank on the one hand, and in the region of Passamoquoddy Bay on the other, by mid- or late March, even by late February in an early spring, with the vicinity of Cape Elizabeth as the site of the earliest flowering. In some years flowerings may develop equally early in the shoal waters on the eastern side of the Gulf, south from Nova Scotia, and over the eastern part of George's Bank, but not until a month or so later in other years, nor until April or even May—according to locality— in the northern coastal belt or over the bowl of the Gulf generally. This regional contrast is illustrated by the charts for March and April (Fig. 2).

In all parts of the Gulf yet studied the vernal augmentation soon results in an increase in the number of cells—chiefly diatoms— so great that by the time the outburst reaches its climax, the number in the water averages at least a thousand times as great as it is in late winter for the Gulf as a whole; in fact multiplication by 63,207-fold has been recorded locally.

The vernal increase in the number of cells per column, after continuing rapid for a period varying locally in duration from a few days to several weeks, then either abruptly ceases, or at least equally abruptly declines in its rate, a stage in the seasonal cycle indicated on graphs of time-abundance distribution by a sudden change in the slope of the curve (e.g., as in Pt. II, Fig. 6 of this report, Lillick, 1940).   This point is named here the "vernal climax," rather than by the more usual designation "vernal peak" as more correctly descriptive of events in regions where high levels of abundance persist for considerable periods.

The vernal climax is reached in the region of Massachusetts Bay by early or mid-April (depending on the year), and equally early on the eastern side of the Gulf generally, including the eastern parts of George's Bank; a week or two later (late April) in the southern part of the western branch of the deep basin; shortly thereafter (late April–early May) in the northern extension of the latter and in the northern coastal belt; latest of all (May or June) in the Bay of Fundy region.   The wide variation in this respect recorded from year to year for Passamoquoddy Bay suggests that the climax may equally be expected to occur as much as three weeks earlier in an early spring than in a late, in most parts of the Gulf as well.

The maximum number of cells recorded has ranged from about 2 billion to about 14 billion per column (Fig. 2), in the parts of the open gulf where counts have been made at or near the vernal climax, in Passamoquoddy Bay and in the lower part of the Bay of Fundy. On this basis no one major subdivision of the Gulf for which information is available can be classed as significantly more productive of phytoplankton than another at the climax season. Nor have we any positive reason to assume any great difference in either direction for George's Bank or for the northern part of the Western Basin, but the inner parts of the Bay of Fundy support a much less abundant flora according to Gran and Braarud's (1935) observations.

Available information for the several years combined shows that the Gulf falls into two subdivisions as regards the phytoplanktonic cycle subsequent to the date at which the vernal climax is reached.   In the one, restricted to the northern coastal belt and to the Bay of Fundy region, a rich flora may persist through June or even into July.   By present indications George's Bank also falls in this category.   In the other subdivision, which includes all other parts of the Gulf, the vernal climax is succeeded by an abrupt impoverishment, most severe in the open bowl outside the 200-meter contour line and from Nova Scotia out across the Eastern Channel where counts were lower in late May (Fig. 2) than at any other season.   The regional contrast between areas rich and barren is also widest in the late spring, when it may be of the order of 174,570 to 1.

A moderate increase in the number of vegetable cells takes place over the bowl generally from July through August following the late spring impoverishment, with more productive flowerings in late summer in the northern coastal belt, in the Bay of Fundy region and on George's Bank; also in September in the Massachusetts Bay region and southward. Whether or not this second flowering results in a definite peak of abundance in any particular region depends on the degree to which aestival impoverishment may have preceded it there.

The second flowering is followed by progressive autumnal impoverishment, the date when this commences depending on the date when the second flowering occurs, which varies regionally and from year to year as described on page 163.   Further depletion through later autumn leads directly to the early winter state outlined above, there being no reason to suppose that active flowerings ever develop even locally in October or in November.

Extreme and mean numbers of cells per column for 1933–1934 were as follows:

PHYTOPLANKTON

Month	Mean	Maximum	Minimum	No. of Stas.	Ratio Maximum to Minimum
December 1933........	1,852,000	3,230,000	80,000	8	40.4 : 1
January 1934.........	6,026,000	12,073,000	1,672,000	10	7 : 1
March 1934..........	29,487,000	117,425,000	550,000	11	213.5 : 1
April–May 1934.......	1,759,767,000	13,965,600,000	80,000	46	174570 : 1
May–June 1934.......	55,347,000	142,100,000	50,000	46	2842 : 1
June 1934............	6,650,000	27,310,000	1,120,000	6	24 : 1
June–July 1933........	21,685,000	198,080,000	800,000	16	247.6 : 1
September 1933.......	37,660,000	283,012,000	2,013,000	11	140 : 1
October 1933.........	7,856,000	18,161,000	649,000	6	28 : 1

The deeper strata of the Gulf below about 80 meters are rich in phosphates (> 100 mgs. per cubic meter) and in nitrates (> 200 mgs. per cubic meter) throughout the year, and values equally high characterize the entire water column up to the surface from December through March. Vernal flowerings do not develop, however, on any large scale until at least 3 months after the stock of phosphates and nitrates has been fully replenished in the photosynthetic zone. Except in special situations such as the turbid waters of the Bay of Fundy, light is also sufficiently intense at the latitudes in question for active photosynthesis throughout the winter, as proved by the occurrence of local flowerings in Cape Cod Bay in December (p. 155). Neither does available evidence suggest that the initiation of general vernal flowerings is dependent on the nutrient increment (of whatever substances) contributed by river freshets, or on the attainment of any particular value of temperature, but that the generally stimulating factor is the attainment of at least a slight degree of stabilization of the water column.

In the year 1934 the vernal outburst of planktonic vegetation reduced the concentration of nitrates in the productive zone to such a degree as to make it probable that the flowerings were largely self-limited in this way. In some years—as in 1932—phosphates also are locally reduced below the minimum requisite for active growth of diatoms. This, however, did not happen in 1934. Thus, by present indications, nitrates are much more likely to act as a generally limiting factor in vernal production than are phosphates. In the Bay of Fundy, however, Gran and Braarud (1935) have maintained that the chief limiting factor is the great turbidity of the water, combined with its turbulence. Available data do not allow quantitative estimates of the effects of grazing by herbivorous members of the zooplankton at this season.

It appears that from midsummer on, the concentration of nitrates tends to increase in the illuminated zone, providing the bases for the second flowerings, but the data are not sufficiently detailed for more than very general statement in this respect. The records for September show local self-limitation of the second flowerings, as of the vernal, by exhaustion of the supply of nitrates (not, however, of phosphates); and it is probable that the effect of grazing is also a major factor in autumnal impoverishment of the phytoplankton.

NUMERICAL DISTRIBUTION OF PLANKTONIC PROTOZOA IN THE UPPER 80
METERS OF WATER

The data for 1933–1934 provide for the first time a picture of the numerical distribution of planktonic protozoa for the Gulf of Maine as a whole at different seasons of the year; previous information regarding this group was confined to a brief discussion by Bigelow (1926), and to counts of zooflagellates and ciliates included under the general heading "Phytoplankton" by Braarud (1934) for the offshore waters in summer, and by Gran and Braarud (1935) for the Bay of Fundy region. We should, however, emphasize that although the counts per column given in the preceding chapter for the planktonic plants include practically the total stock of the latter because of their limitation to the illuminated zone, this does not necessarily apply to the protozoa which, on occasion, may have been present in significant numbers below 80 meters [21] in the deeper parts of the gulf, as they certainly were at several of the Fundian stations (Gran and Braarud, 1935, tables). Hence the counts are minimal; complete enumeration for the entire water column, surface to bottom, would be larger by an unknown amount.

The collections of 1933–1934 include Heliozoa, Radiolaria, Foraminiferae, Ciliatae, Zoomastigodae, a few members of the Sarcodina (mainly Amoebae), which were moderately abundant in the spring, and others not identified, with the Ciliatae forming the larger part. The genera most commonly represented are Strombidium, Mesodinium, Parafavella, Tintinnus, Tintinnopsis, Stenosemella, and Laboea. Gran (1933), Braarud (1934) and Gran and Braarud (1935) also give counts of various other ciliates. Foraminiferae were extremely rare, not more than a dozen individuals having been detected in the entire survey; and Radiolaria were not found at all in the year 1933–1934, although Acanthometron was abundant near Cape Ann and in the western basin in August 1914, comparatively plentiful at scattered localities in September 1915, and occasional west and south of Nova Scotia in April 1920 (Bigelow, 1926, p. 460). Few zooflagellates, other than Zoomastigodae, were observed by us. Bodo is, however, recorded in great numbers at two stations in the open Gulf for July 1933 by Braarud (1934), hence it is probable that these delicate and easily destroyed forms were generally abundant as Lackey (1936) found them near Woods Hole, but that they had either distintegrated before the collections of 1933–1934 were examined, or were damaged beyond recognition by the methods used.

The protozoa have not been identified as to species; taxonomic discussion of the group is therefore omitted here. For such in the Bay of Fundy, see Gran and Braarud (1935).

The protozoan counts presented on Figs. 9–10 are summed up in the following table.

Month	Year	Mean	Maximum	Minimum	No. of Station
December.............	1933	435,000	950,000	90,000	8
January..............	1934	291,000	891,000	0	10
March...............	1934	1,405,000	10,780,000	0	11
April–May...........	1934	630,000	1,380,000	20,000	10
Late May............	1934	320,000	56,000	50,000	6
June................	1934	470,000	2,000,000	0	6
June–July...........	1934	1,208,000	3,920,000	0	16
September...........	1933	489,000	1,595,000	0	11
October.............	1933	474,000	1,265,000	55,000	6

[21] The deepest sampling level for the cruises of 1933–1934.

If the seasonal succession as it occurred in 1933–1934 be representative, the protozoa are at their minimum in the Gulf in January, richest then close to Cape Cod (910,000) and on the eastern part of George's Bank (950,000) (Fig. 9). To illustrate the sharpness of the contrast that may then exist within short distances, we may cite a count of 493,000 off Port-

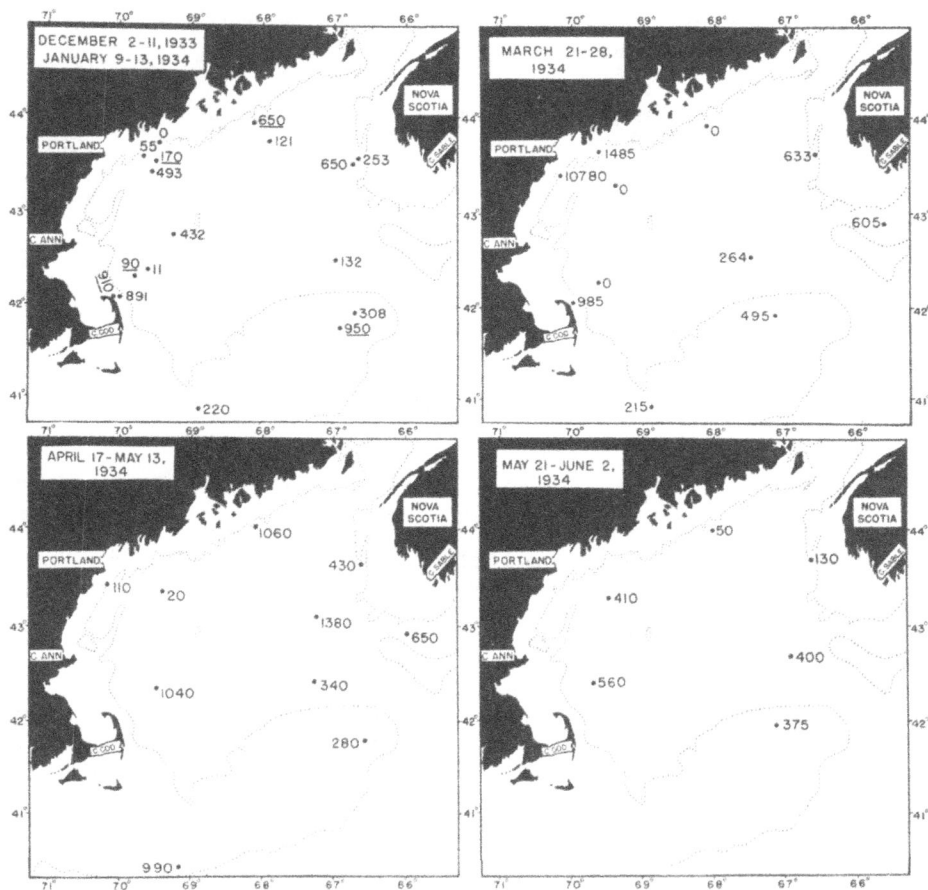

FIG. 9. Numbers of protozoa, of groups sampled (p. 182) in thousands, per 0.1 square meter of sea surface. Values for December underlined.

land, whereas none at all were detected close to the coast near Seguin Island only about 20 miles distant. In the year in question the protozoa had multiplied by March to their yearly maximum (both as to average numbers and as to maxima), when the average was about 4.8 times as great as it had been two months before and the maximum count 12 times greater, though the minimum was the same (0 in each case). These averages are, however, deceptive if taken at face value, for actually the ostensible increase in richness resulted almost entirely from the development of a center of abundance near Portland, where a local swarming of Strombidium raised the March counts (2 stations, average, 6,132,000) to about 22 times as great as those for January (2 stations, average, 275,000), whereas in no other part of the Gulf was the count for either month as much as three times as great as for the other (cf. Figs. 9–10).

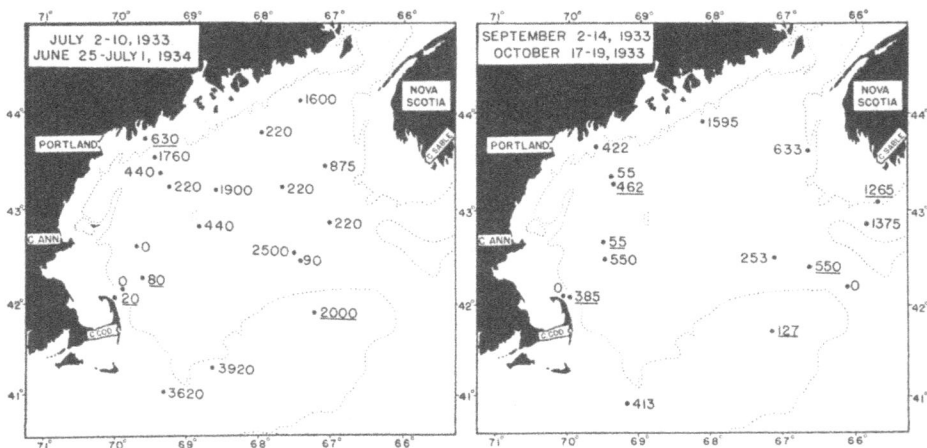

FIG. 10.   Numbers of protozoa of groups sampled (p. 182) in thousands per 0.1 square meter of sea surface. Values for June–July, 1934, and October, 1933, underlined.

By late April (Fig. 9) the count for the locality that had been so rich in March had fallen to about 110,000 cells per column, or to about 0.18 of the March value; this was, in fact, the region of greatest paucity at the time of the April–May cruise, when Protozoa were the most abundant at widely separated localities in the southwestern, southeastern, and northeastern parts of the basin (Fig. 9).   The correspondence between the late April counts, near the 100-meter line off Mt. Desert Island for 1934 (1,000,000, Fig. 9), and for 1932 (1,210,000) calculated from Gran and Braarud's (1935, table, p. 440) data is very close.   However, an April count of 1,630,000 in the northwestern part of the basin for 1932 (Gran and Braarud, 1935) contrasted with 20,000 only at the same general locality in 1934, shows that information in much greater detail will be needed before the significant range of annual variation can be evaluated for this part of the Gulf or for any other.

In the year 1934 vernal enrichment appears to have been confined to rather definitely circumscribed centers, and no great change is indicated in abundance from winter through the early spring on the Nova Scotian side of the Gulf where the recorded counts for March, April and early May varied only between about 0.5 times and 2.5 times as great as those for January (cf. Fig. 9).   The average number of protozoa for the Gulf as a whole (320,000) had again declined by the end of May (Fig. 9) to about what it had been in January [22]

[22] Data are lacking for the region of Massachusetts Bay.

(291,000). And the fact that the late May average for the northwestern part of the Gulf was at least of the same order of magnitude in 1932 (about 107,000, 3 stations) as in 1934 (230,000, 2 stations, Fig. 9), makes it likely that general impoverishment at this time of year is a normal stage in the seasonal cycle.

In 1934 an 8-fold increase in the number of protozoa took place on the eastern part of George's Bank at some time between the dates of the late May and late June cruises; also, in the coastal belt near Portland there were about 6 times as many protozoa on June 29 as there had been on April 19 (no data there for May). And while the protozoan population in this last locality was very sparse in late June of 1932 (40,000 from Gran and Braarud's table; no data for that April), the sequence recorded in 1934 appears to have presaged a second and general enrichment in the summer, since the counts for early July of 1933 were more than 3 times as great as those for the spring of 1934, wherever pairs of stations are available for comparison, except near Mt. Desert Island, and twice as great even there (cf. Figs. 9, 10). The average was in fact nearly as great in June–July as for March, the maximum (3,920,000) nearly as great, and the minimum the same (0), while the second richest count for July (3.6 million in the South Channel) ranks considerably above the maximum for any other month except March, so far as the open Gulf is concerned. And the general distribution shown on Fig. 10 as contrasted with Fig. 9 is evidence that the aestival enrichment was much more widespread than the vernal.

Braarud's (1934) counts for two July stations (southeastern part of basin and eastern George's Bank) are especially interesting, not only because they provide the only available evidence as to the abundance of zooflagellates in the open Gulf at any season, but also because they prove that these may outnumber all other protozoa combined in midsummer, for he lists no less than 10,592,000 per column of one genus (Bodo) alone at the one station, and 5,112,000 at the other. But we have no indication as yet whether this represents a generally high level of abundance at that season or whether the two stations in question chanced to sample particularly rich centers for this particular flagellate genus. Acanthometron swarmed off Cape Ann in late August of 1914 (Bigelow, 1917, Fig. 97; 1926, p. 460), but this radiolarian is so large that the numbers concerned were probably not as great as were the July counts for other protozoa (Fig. 10) in 1933. Present indications therefore are that protozoa of all sorts combined usually do not multiply to more than about 30 million per column anywhere in the Gulf during the summer.

Numerical data for the open Gulf for August are confined to counts by Gran (1933) corresponding to between 152,000 and 528,000 per column in the deep basin and 96,000 on the eastern part of George's Bank in 1932, and by Gran and Braarud (1935, Tables) of 160,000 and 296,000 off Seguin Island and off Mt. Desert Island. These suggest about the same general level of abundance in the open Gulf for August as for July; the failure to encounter any very rich centers in August may have been simply a matter of chance. The highest counts for July (Fig. 10) and August are so widely scattered that no one subdivision of the Gulf can be classed as definitely more productive than another at that season—at least on the basis of present evidence. The fact, however, that August counts for the Bay of Fundy in 1932 by Gran and Braarud (1935, Tables) ranged from 1,208,000 to 3,856,000 per column (average 2,314,000) contrasted with a maximum of 528,000 and an average of only 205,000 in the open waters to the southward, is evidence that the summer stock of protozoa is considerably richer within the Bay than it is outside the latter, even though no center of abundance has yet been encountered in the Bay to rival the July concentration recorded in the southeastern part of the Gulf a year later.   It is not likely that this regional contrast between the Bay and the open Gulf is established before mid-summer since counts equalling 1,350,000 protozoa per column in Passamoquoddy Bay and 104,000 in the Bay of Fundy for late April of 1932, 248,000 for the latter a month later, and 80,000, 816,000 and 990,000 (3 stations) for June of 1932 are of the same general range of magnitude as the late April–June counts in one part of the Gulf or another for 1934 (Fig. 9).

In the year 1933 comparatively large numbers of protozoa (1,595,000–1,375,000) were again recorded in the Eastern Channel between Brown's and George's Banks and on the northern side of the Gulf off Mt. Desert Island in September (Fig. 10), in line with which last is an earlier record of abundant radiolarians near Penobscot Bay at the same season (Bigelow, 1926, Fig. 132).   Gran and Braarud (1935) also had a very high count at one station within the Bay of Fundy in that month of 1932 (equalling about 2 million per column), with two others in September of 1931, i.e., one of about 7 million within Passamoquoddy Bay, the other of about 1 million off its mouth.   Elsewhere, however, the September counts have averaged considerably below those for the preceding two months (whether in the open Gulf or in the Bay); and by October of 1933 (Fig. 10) the numbers per column— mean, maximum and minimum—had fallen to about as low a level as they were in December, though with considerable divergences between individual stations (cf. Figs. 9, 10).   The record does not point to any significant change during the early winter—or indeed until the vernal enrichment develops—for although the general average was somewhat larger for December (Table, p. 182) than for January, the maxima were not far apart.

SUMMARY

Planktonic protozoa are at their numerical minimum in the Gulf from mid-autumn through the winter.   No widespread important secular fluctuations in abundance are indicated for this part of the year, but the regional variation was wider in January than in any other month, with the most abundant winter population (900,000–1,000,000) in shoal water close to Cape Cod and on the eastern part of George's Bank.   In the one year of record a very rich population developed locally in the coastal belt near Portland in March, the result of swarming of one form (Fig. 9).   But apart from the two stations directly affected, the average was not greater for March (about 355,000) than for December–January combined (366,000); and the swarm in question was not only confined to a very small area, but did not endure more than 3 weeks, its disappearance leaving the population for the Gulf as a whole not significantly richer in April than during the preceding winter except at isolated stations shown on figure 9.   Combination of data for the two years 1933– 1934 points, however, to considerable increase in the population from May through June, resulting in counts from 2 to 20 times as high at corresponding stations in early July as in January, with the notable exception of two localities near the tip of Cape Cod, where protozoa appear to have disappeared entirely from the water if zooflagellates be left out of account.   Even if these stations be included in the calculation, the June–July average (about 1,208,000) is 4 times that for January, the maximum about 2 times, though the minimum (0) was the same (see Table, p. 182).   Two very rich catches of the genus Bodo have also been recorded in July (p. 185).

The counts for 1933–1934 added to various pieces of information for previous years mark mid-summer as the season of maximum for protozoa over the Gulf as a whole, and point to some 4 million per column as approximately the maximum to be expected then. No notably rich counts have been recorded in August; the general level of abundance for protozoa appears, however, to be about the same for that month as for July.

The distribution of the centers of maximum abundance does not point to any one major subdivision of the open gulf—deeper or shoal—as characteristically more productive of protozoa than another at the season of the aestival maximum, but the group averages considerably more abundant in the inner parts of the Bay of Fundy in summer than elsewhere within the region studied.

Locally protozoa may continue abundant into September both in the open Gulf (maximum, 1,595,000) and in the Bay of Fundy (maximum, about 7 million). The general level of abundance averages considerably lower then, however, than in August or July, and by October the counts (both mean and maximum) fall back, roughly, to the December level. Nor have we any grounds for supposing that centers of active production develop anywhere during the late autumn or early winter. The next to the smallest count for protozoa (11,000 per column) was only about 1/8 as great as the smallest for phytoplankton; the largest for protozoa (10,780,000) about 1/1295 as great as the largest for phytoplankton; the grand average for protozoa only about 1/337 that for phytoplankton. These contrasts sufficiently emphasize the paucity of the waters of the Gulf of Maine in unicellular animals as contrasted with unicellular plants.

BIBLIOGRAPHY

ATKINS, W. R. G.
  1925 Seasonal Changes in the Phosphate Content of Sea Water in relation to the Growth of the Algal Plankton during 1923 and 1924. Jour. Mar. Biol. Assoc., N. S., Vol. 13, pp. 700–720, 8 textfigs.
  1927 The Phosphate Content of Sea Water in relation to the Growth of the Algal Plankton. III. Jour. Mar. Biol. Assoc., N. S., Vol. 14, pp. 447–467, 5 textfigs.
  1928 Seasonal Variations in the Phosphate and Silicate Content of Sea Water during 1926 and 1927 in Relation to the Phytoplankton Crop. Jour. Mar. Biol. Assoc., N. S., Vol. 15, pp. 191–205.
  1929 Seasonal Variations in the Phosphate and Silicate Content of Sea Water in Relation to the Phytoplankton Crop. Part V. November 1927 to April 1929, Compared with Earlier Years from 1923. Jour. Mar. Biol. Assoc., N. S., Vol. 16, pp. 821–852, 12 textfigs.

BAILEY, L. W.
  1910 Marine and Estuarine Diatoms of the New Brunswick Coasts. Bull. Nat. Hist. Soc. New Brunswick, Vol. 6, pp. 219–236.
  1912 Some recent diatoms, freshwater and marine, from the vicinity of the biological station, St. Andrews, N. B., August, 20–30, 1909. Contrib. Canadian Biol., 1906–1910, pp. 243–264, 2 pls.
  1915 The plankton diatoms of the Bay of Fundy. Contrib. Canadian Biol., 1911–1914, Fasc. I. Marine Biology, Suppl. 47th Ann. Rep. Dept. of Mar. and Fish. Invest., Fisheries Branch, pp. 11–23, 3 pls.
  1917 Notes on the phytoplankton of the Bay of Fundy and Passamoquoddy Bay. Contrib. Canadian Biol., 1915–1916. Suppl., 6th Ann. Rep., Dept. of Naval Service, Fisheries Branch, pp. 93–107.
  1924 An Annotated Catalogue of the Diatoms of Canada, showing their Geographical Distribution. Contrib. Canadian Biol., N. S., Vol. 11, pp. 31–37.

BAILEY, L. W., AND MACKAY, A. H.
  1921   The Diatoms of Canada. Contrib. Canadian Biol., 1918–1920, Dept. Naval Service, pp. 115–124.

BIGELOW, H. B.
  1914   Explorations in the Gulf of Maine, July and August, 1912, by the U. S. Fisheries schooner *Grampus*. Oceanography and notes on the plankton. Bull. Mus. Comp. Zool., Vol. 58, No. 2, pp. 29–147, 38 textfigs., 9 pls.
  1914a   Oceanography and plankton of Massachusetts Bay and adjacent waters, November, 1912–May, 1913. Bull. Mus. Comp. Zool., Vol. 58, No. 10, pp. 385–419, 7 textfigs., 1 pl.
  1915   Exploration of the coast water between Nova Scotia and Chesapeake Bay, July and August, 1913, by the U. S. Fisheries schooner *Grampus*. Oceanography and plankton. Bull. Mus. Comp. Zool., Vol. 59, No. 4, pp. 149–359, 82 textfigs., 2 pls.
  1917   Explorations of the coast water between Cape Cod and Halifax in 1914 and 1915, by the U. S. Fisheries schooner *Grampus*. Oceanography and plankton. Bull. Mus. Comp. Zool., Vol. 61, No. 8, pp. 161–357, 100 textfigs., 2 pls.
  1922   Exploration of the coastal water off the northeastern United States in 1916 by the U. S. Fisheries schooner *Grampus*. Bull. Mus. Comp. Zool., Vol. 65, No. 5, pp. 87–188, 53 textfigs.
  1926   Plankton of the offshore waters of the Gulf of Maine. Bull. U. S. Bureau of Fisheries, Vol. 40, pp. 1–509, 134 textfigs.

BRAARUD, TRYGVE
  1934   A note on the phytoplankton of the Gulf of Maine in the summer of 1933. Biol. Bull., Vol. 67, pp. 76–82.

BRAARUD, TRYGVE, AND KLEM, ALF
  1931   Hydrographical and chemical investigations in the coastal waters off Møre and in the Romsdalsfjord. Hvalrådets Skrift. Norske Vidensk.-Akademi i Oslo, No. 1, pp. 1–88, 21 textfigs.

BURKHOLDER, P. R.
  1933   A Study of the Phytoplankton of Frenchmans Bay and Penobscot Bay, Maine. Internat. Rev. d. ges. Hydrobiol. u. Hydrogr., Vol. 28, pp. 262–284, 6 textfigs.

COOPER, L. H. N.
  1933   Chemical Constituents of Biological Importance in the English Channel, November, 1930, to January, 1932. Part 1. Phosphate, silicate, nitrate, nitrite, ammonia. Jour. Mar. Biol. Assoc., N. S., Vol. 18, pp. 677–728, 15 textfigs.
  1933a   Chemical Constituents of Biological Importance in the English Channel. III. June–December, 1932. Phosphate, Silicate, Nitrate, Hydrogen Ion Concentration, with a Comparison with Wind Records. Jour. Mar. Biol. Assoc., Vol. 19, pp. 55–62, 3 textfigs.
  1937   On the ratio of nitrogen to phosphorus in the sea. Jour. Mar. Biol. Assoc., N. S., Vol. 22, pp. 177–182.
  1937a   The nitrogen cycle in the sea. Jour. Mar. Biol. Assoc., N. S., Vol. 22, pp. 183–204, 2 textfigs.
  1938.   Phosphate in the English Channel, 1933–8, with a comparison with earlier years, 1916 and 1923–32. Jour. Mar. Biol. Assoc., N. S., Vol. 23, pp. 181–195, 1 textfig.

DAVIDSON, V. M.
  1934   Fluctuations in the Abundance of planktonic diatoms in the Passamoquoddy Region, New Brunswick, from 1924 to 1931. Contrib. Canadian Biol. and Fish., N. S., Vol. 8, No. 28, pp. 357–407, 33 textfigs.

ERCEGOVIĆ, A.
  1940   Weitere Untersuchungen über einige hydrographische Verhältnisse und über die Phytoplankton-production den Gewässern der östlichen Mitteladria. Acta Adriatica Instituti Oceanographia Split (Jugoslavija), Vol. II, No. 3, pp. 95–134, 8 textfigs.

FISH, CHARLES J.
  1925   Seasonal distribution of the plankton of the Woods Hole region. Bull. U. S. Bureau of Fisheries, Vol. 41, pp. 91–179, 81 textfigs.

FLEMING, RICHARD H.
1939 The Control of Diatom Populations by Grazing. Jour. du Cons., Cons. Perm. Int. Expl. de la Mer, Vol. 14, No. 2, pp. 210–227, 5 textfigs.

FØYN, BERGITHE RUUD
1929 Investigation of the phytoplankton at Lofoten, March–April, 1922–1927. Skrift. Norske Vidensk.-Akademi i Oslo, 1928, I Matem.-naturvidenskap. Klass., Vol. 1, pp. 1–71, 15 textfigs.

FRITZ, CLARA W.
1921 Plankton Diatoms, their Distribution and Bathymetric Range in St. Andrews Waters. Contrib. Canadian Biol., 1918–1920, Dept. Naval Service, pp. 49–62, 3 pls.

GESSNER, FRITZ
1933 Phosphat, Nitrat und Planktongehalt im Arkonabecken. Ein Beitrag zur Produktionsbiologie der Ostsee. Jour. du Cons., Cons. Perm. Int. Expl. de la Mer, Vol. 8, pp. 181–194, 5 textfigs.

GILLAM, A. E., EL RIDI, M. S., AND WIMPENNY, R. S.
1939 The seasonal variation in biological composition of certain plankton samples from the north sea in relation to their content of vitamin A, carotenoids, chlorophyll, and total fatty matter. Jour. Exper. Biol., Vol. 16, No. 1, pp. 71–88, 7 textfigs.

GRAN, H. H.
1930 The Spring growth of the plankton at Møre in 1928–1929 and at Lofoten in 1929 in relation to its limiting factors. Skrift. Norske Vidensk.-Akademi i Oslo, I Matem. naturvidenskap. Kl. for 1930, No. 5, 77 pp., 7 textfigs.
1933 Studies on the biology and the chemistry of the Gulf of Maine. II. Distribution of Phytoplankton in August, 1932. Biol. Bull., Vol. 64, No. 2, pp. 159–182.

GRAN, H. H., AND BRAARUD, TRYGVE
1935 A Quantitative Study of the Phytoplankton in the Bay of Fundy and the Gulf of Maine (including Observations on Hydrography, Chemistry and Turbidity). Jour. Biol. Board of Canada, Vol. 1, pp. 279–467, 69 textfigs.

GRAN, H. H., AND THOMPSON, THOMAS G.'
1930 The Diatoms and the Physical and Chemical Conditions of the Sea Water of the San Juan Archipelago. Publ. Puget Sound Biol. Sta., University of Washington, Vol. 7, pp. 169–204, 9 textfigs.

HARVEY, H. W.
1926 Nitrate in the Sea. Jour. Mar. Biol. Assoc., N. S., Vol. 14, pp. 71–88, 3 textfigs.
1928 Nitrate in the Sea. II. Jour. Mar. Biol. Assoc., N. S., Vol. 15, pp. 183–190, 3 textfigs.
1928a Biological Chemistry and Physics of Sea Water. 194 pp., 65 textfigs., The University Press, Cambridge, England.
1934 Measurement of Phytoplankton Population. Jour. Mar. Biol. Assoc., N. S., Vol. 19, pp. 761–773, 9 textfigs.
1934a Annual Variation of Planktonic Vegetation, 1933. Jour. Mar. Biol. Assoc., N. S., Vol. 19, pp. 775–792, 5 textfigs.
1939 Substances controlling the growth of a diatom. Jour. Mar. Biol. Assoc., N. S., Vol. 23, No. 2, pp. 499–520.

HARVEY, H. W., COOPER, L. H. N., LEBOUR, M. V., AND RUSSELL, F. S.
1935 Plankton Production and its Control. Jour. Mar. Biol. Assoc., Vol. 20, pp. 407–441, 16 textfigs.

KETCHUM, BOSTWICK H.
1939 The absorption of phosphate and nitrate by illuminated cultures of Nitzschia Closterium. Am. Jour. Bot., Vol. 26, pp. 399–407, 4 textfigs.
1939a The development and restoration of deficiencies in the phosphorus and nitrogen composition of unicellular plants. Jour. Cell. and Comp. Physiol., Vol. 13, pp. 373–381.

KIMBALL, HERBERT H.
1928 Amount of solar radiation that reaches the surface of the earth on the land and on the sea, and methods by which it is measured. Monthly Weather Review, Vol. 56, No. 10, pp. 393–398, 7 textfigs.
KREPS, E., AND VERJBINSKAYA, N.
1930 Seasonal Changes in the Phosphate and Nitrate Content and in Hydrogen Ion Concentration in the Barents Sea. Jour. du. Cons., Cons. Perm. Int. Expl. de la Mer, Vol. 5, No. 3, pp. 329–346, 8 textfigs.
KREY, JOHANNES
1939 Die Bestimmung des Chlorophylls in Meerwasser-Schöpfproben. Jour. du Cons., Cons. Perm. Int. Expl. de la Mer, Vol. 14, No. 2, pp. 201–209, 4 textfigs.
LACKEY, J. B.
1936 Occurrence and distribution of the marine protozoan species in the Woods Hole area. Biol. Bull., Vol. 70, pp. 264–278.
LILLICK, L. C.
1938 Preliminary report of the phytoplankton of the Gulf of Maine. Amer. Midl. Nat., Vol. 20, pp. 624–640, 1 textfig.
1940 Phytoplankton and Planktonic Protozoa of the Offshore Waters of the Gulf of Maine. Part II. Qualitative Composition of the Planktonic Flora. Trans. Am. Philos. Soc., Vol. XXXI, Pt. III, pp. 149–191, 13 textfigs., 4 tables.
MARSHALL, S. M., AND ORR, A. P.
1926 The Relation of the Plankton to some Chemical and Physical Factors in the Clyde Sea Area. Jour. Mar. Biol. Assoc., N. S., Vol. 14, pp. 837–868, 9 textfigs., 10 pls.
MOBERG, ERIK G.
1928 The interrelation between diatoms, their chemical environment, and upwelling water in the sea, off the coast of southern California. Proc. Nat. Acad. Sci., Vol. 14, pp. 511–518, 1 textfig.
McMURRICH, J. P.
1917 The Winter plankton in the neighborhood of St. Andrews, 1914–1915. Contrib. Canadian Biol. 1915–1916. Suppl. 6, Ann. Rep. Dept. Naval Service, Fisheries Branch, pp. 1–8.
NATHANSOHN, A.
1910 Études hydrobiologiques d'après les recherches faites à bord de l'"Eider" au large de Monaco de janvier à juillet 1909. Ann. Instit. Oceanogr., Monaco, Vol. 1, No. 5, pp. 1–27, 3 pls.
PHIFER, LYMAN D., AND THOMPSON, THOMAS G.
1937 Seasonal variations in the surface waters of San Juan Channel during the five year period, January 1931 to December 30, 1935. Jour. Mar. Res., Sears Found., Vol. 1, No. 1, pp. 34–59, textfigs. 12–18.
RAKESTRAW, N. W.
1933 Studies on the biology and chemistry of the Gulf of Maine. I. Chemistry of the Waters of the Gulf of Maine in August, 1932. Biol. Bull., Vol. 64, No. 2, pp. 149–158, 4 textfigs.
1936 The occurrence and significance of nitrite in the sea. Biol. Bull., Vol. 71, pp. 133–167, 12 textfigs.
RAKESTRAW, N. W., AND SMITH, H. P.
1937 A contribution to the chemistry of the Caribbean and Cayman Seas. Bull. Bingham Oceanographic Coll., Vol. 6, pp. 1–41, 32 textfigs.
REDFIELD, ALFRED C., AND KEYS, ANCEL B.
1938 The distribution of ammonia in the waters of the Gulf of Maine. Biol. Bull., Vol. 74, No. 1, pp. 83–92, 6 textfigs.
REDFIELD, ALFRED C., SMITH, HOMER P., AND KETCHUM, BOSTWICK
1937 The cycle of organic phosphorus in the Gulf of Maine. Biol. Bull., Vol. 73, No. 3, pp. 421–443, 4 textfigs.

RILEY, GORDON A.

1939   Correlations in Aquatic Ecology with an example of their applications to problems of plankton productivity. Jour. Mar. Res., Sears Found. Mar. Res., Vol. 2, No. 1, pp. 56–73, textfigs. 14–16.

1939a   Plankton Studies. II. The Western North Atlantic, May–June 1939. Jour. Mar. Res., Sears Found. Mar. Res., Vol. 2, No. 2, pp. 145–162, textfigs. 49–51.

RUUD, JOHAN T.

1930   Nitrates and Phosphates in the Southern Seas. Jour. du Cons., Cons. Perm. Int. Expl. de la Mer, Vol. 5, No. 3, pp. 347–360, 5 textfigs.

STEEMANN NIELSEN, E.

1938   Über die Anwendung von Netzfängen bei quantitativen Phytoplanktonuntersuchungen. Jour. du Cons., Cons. Perm. Int. Expl. de la Mer., Vol. 13, No. 2, pp. 197–205, 1 textfig.

STEEMANN NIELSEN, E., AND V. BRAND, TH.

1934   Quantitative Zentrifugenmethoden zur Planktonbestimmung. Rapp. et Proc. Verb., Cons. Perm. Int. Explor. de la Mer, Vol. 89, Pt. 3, appendices, pp. 99–100.

UNTERMÖHL, H.

1931.   Neue Wege in der quantitativen Erfassung des Planktons (Mit besonderer Berücksichtigung des ultraplanktons). Verhandl. Int. Vereinig. f. Theoretische u. angewandte Limnol., Vol. 5, pp. 567–596, 4 textfigs.

PHYTOPLANKTON AND PLANKTONIC PROTOZOA OF THE OFFSHORE WATERS OF THE GULF OF MAINE

Part II—Qualitative Composition of the Planktonic Flora [1]

By Lois C. Lillick

CONTENTS

ABSTRACT

The scanty winter flora of the Gulf of Maine (usually dominated by Coscinodiscus, by Ceratium, or by other small peridinians) is succeeded by the vernal outbursts of diatoms and a decrease in the abundance of peridinians. This flowering results chiefly from the rapid multiplication of Thalassiosira. The outburst starts first in the western coastal belt (February–early April); soon thereafter (March–mid-April) in the coastal and bank waters west and south of Nova Scotia, possibly also on George's Bank. The area of the Thalassiosira flowering expands northward along shore on both sides of the gulf, also southward and offshore into the eastern basin and to a certain extent into the western basin and over the eastern part of George's Bank. Active multiplication is briefest (2–4 weeks) in the eastern and southeastern parts of the gulf generally, longest in the northern coastal belt. After reaching its peak, Thalassiosira falls within a few days to an insignificant rank in the flora.

Accompanying the eclipse of Thalassiosira, there is a flowering of Chaetoceros, which lasts from 4–6 weeks over the gulf generally (i.e., until late April–May). Once its peak of abundance is passed, Chaetoceros declines as abruptly in abundance as Thalassiosira.

[1] Contribution No. 262 from the Woods Hole Oceanographic Institution; Papers from the Department of Botany of the University of Michigan No. 707.

Diatoms (though at times outnumbered by Pontosphaera) continue to dominate the flora on George's Bank throughout the summer and in some years on the northern coastal belt as well. In other years, the latter belt is dominated by a sparse peridinian flora (chiefly Ceratium) accompanied by coccolithophorids, as is the coastal belt farther south and the basin of the gulf generally, at some time between the end of May and the beginning of July.

Second flowerings of diatoms develop locally during the late summer, Rhizosolenia being frequently responsible for these in the shoaler waters including George's Bank, although Chaetoceros may have a distinct second flowering in the northern coastal belt. *Guinardia flaccida* is also an important component of the late summer flowering, especially on George's Bank, and *Sceletonema costatum* inshore. Peridinians, however, persist in undiminished numbers throughout the late summer and coccoliths are also at their maximum.

During the autumn the diatom flora is most varied and the regional contrast widest qualitatively, for with the end of active diatom proliferation, the flora gradually assumes the early winter state.

The seasonal cycle is essentially the same in the Bay of Fundy, except that Biddulphia is more important in the vernal flowering, that the vernal peaks of Thalassiosira and Chaetoceros develop later and endure longer there, and that tychopelagic and neritic diatoms are relatively more important in summer, peridinians less so.

In the Woods Hole region, the maximum flowering takes place several months earlier in the season and is chiefly of species that are of little importance in the winter-spring flora of the gulf.

The records of 1933–1934 corroborate observations that the phytoplankton cycle in the Gulf of Maine parallels that of the boreal waters of Europe in time-sequence, differing only in the association of species concerned.

INTRODUCTION

The following discussion is limited to the species that are the more important numerically in the Gulf of Maine.[2] The seasonal distribution of others of which only small numbers have been recorded, and of various tychopelagic and brackish water forms, which do not persist in the plankton of the open gulf, is presented in Tables I–III.

THE GULF OF MAINE PROPER [3]

Earlier studies (Bigelow, 1926; Gran, 1933; Braarud, 1934; and Gran and Braarud, 1935) had already shown that the phytoplankton of the Gulf of Maine closely resembles that of the Norwegian coast and of the Skagerak in its general composition, with the same boreal species playing the major rôles from season to season. Diatoms, of which 130 species are recorded from the plankton of the gulf, constitute by far the major portion, peridinians and coccolithophorids ranking next. Phaeocystis may be dominant locally (but briefly), silicoflagellates are widespread, though seldom prominent, and the green alga Halosphaera has been recorded at many stations, though always in small numbers. The samples have also contained a few representatives of the Myxophyceae, genera Merismopedia and Gloeocapsa, as well as many small unidentified pigmented forms classed together here for convenience as "flagellates." No doubt some of these are members of the unarmored Dinophyceae, others are the spores of phytoplankton, and still others are marine green flagellates (p. 232). Fungi (spores and mycelia) were also present in small numbers in most of the collections, most abundantly at the deeper levels, but no estimate of their abundance has been attempted. Most of the spores belong to the common genera Al-

---

[2] For a preliminary report, including charts of numerical abundance of individual species, see Lillick (1938).

[3] For a chart of the Gulf of Maine, see Part I of this report (Bigelow, Lillick, and Sears, 1940).

TABLE I

Seasonal Distribution of the Bacillariophyceae of the Gulf of Maine

Species	Months											
	J	F	M	A	M	J	J	A	S	O	N	D
Achnanthes taeniata Grun.				×	×							
Actinocyclus Ehrenbergi Ralfs					×	×	×	×	×			
Actinoptychus undulatus (Bail.) Ralfs	×	×	×	×	×	×	×	×	×	×	×	×
Amphiprora alata Kütz.									×			
Asterionella gracillima (Hantzsch) Heib.					×							
A. japonica Cleve			×	×	×	×	×	×	×			
A. kariana Grun.				×	×	×	×	×	×			
Bacteriastrum fragilis Gran					×							
Biddulphia aurita (Lyngb.) Brèb. & God.	×	×	×	×	×	×	×	×	×	×	×	×
B. regia (Schulze) Ostf.									×			
Cerataulina Bergoni Perag.	×			×	×	×	×	×	×			
Chaetoceros spp.	×	×	×	×	×	×	×	×	×	×	×	×
Ch. affinis Laud.				×	×	×	×	×	×			
Ch. affinis var. *Willei* (Gran) Hust.					×							
Ch. atlanticus Cleve		×	×	×	×	×	×	×				
Ch. borealis Bail.	×	×	×	×	×	×	×	×	×	×	×	×
Ch. brevis Schütt				×				×	×			
Ch. ceratosporum Ostf.					×							
Ch. cinctus Gran						×	×	×				
Ch. compressus Laud.				×	×	×	×	×	×			
Ch. concavicornis Mang.				×	×							
Ch. constrictus Gran				×	×	×	×	×	×	×		
Ch. convolutus Castr.		×	×	×	×	×	×	×	×			
Ch. crinitus Schütt							×	×	×			
Ch. curvisetus Cleve					×	×	×					
Ch. danicus Cleve							×	×	×			
Ch. debilis Cleve				×	×	×	×	×	×	×		
Ch. decipiens Cleve	×	×	×	×	×	×	×	×	×	×	×	×
Ch. densus Cleve				×	×	×	×	×	×			
Ch. didymus Ehr.				×	×	×	×	×	×			
Ch. furcellatus Bail.					×	×	×					
Ch. gracilis Schütt							×					
Ch. laciniosus Schütt				×	×	×	×	×	×	×		
Ch. lorenzianus Grun.					×							
Ch. pelagicus Cleve												
Ch. radicans Schütt					×	×	×	×	×			
Ch. seiracanthus Gran					×							
Ch. similis Cleve							×					
Ch. simplex Ostf.							×	×	×	×		
Ch. socialis Laud.			×	×	×							
Ch. subsecundus (Grun.) Hust.			×	×	×	×	×	×	×			
Ch. subtilis Cleve									×			
Ch. teres Cleve			×	×	×	×	×	×				
Ch. Wighami Brightw.												
Cocconeis sp.			×									
C. placentula Ehr.	×	×	×	×	×	×	×	×	×	×	×	×

TABLE I (*Continued*)

Species	Months											
	J	F	M	A	M	J	J	A	S	O	N	D
C. Scutellum Ehr.			×				×	×				
Corethron hystrix Hensen	×		×	×	×	×	×	×				×
Coscinodiscus spp.	×	×	×	×	×	×	×	×	×	×	×	×
C. argus Ehr.		×										
C. asteromphalus Ehr.			×	×	×	×	×	×	×	×		
C. centralis Ehr.	×	×	×	×	×	×	×	×	×	×	×	×
C. cinctus Kütz.					×				×			
C. concinnus W. Smith			×	×	×	×	×	×				
C. curvatulus Grun.			×	×	×							
C. excentricus Ehr.	×	×	×	×	×		×	×	×	×	×	×
C. lineatus Ehr.					×			×	×			
C. osculis-iridis Ehr.				×	×			×	×	×		
C. radiatus Ehr.			×	×	×		×	×	×			
C. stellaris Roper							×					
Coscinosira Oestrupi Ostf.				×	×	×	×	×	×			
C. polychorda Gran				×	×	×	×	×	×			
Cyclotella sp.	×											
Dactyliosolen mediterraneus Perag.							×					
Detonula confervacea (Cleve) Gran	×		×	×	×	×	×					
Ditylium Brightwelli (West) Grun		×	×	×	×	×	×	×	×	×		
Endictya oceanica Ehr.				×	×	×						
Eucampia recta Gran & Braarud							×	×				
E. zoodiacus Ehr.		×		×	×	×	×	×				
Eunotia arcus Ehr.				×								
Fragilaria sp.				×								
F. cylindrus Grun.				×			×	×				
F. oceanica Cleve				×								
Grammatophora marina (Lyngb.) Kütz.							×	×	×			×
Guinardia flaccida (Castr.) Perag.	×	×				×	×	×	×			
Hemiaulus Hauckii Grun.							×					
Leptocylindrus danicus Cleve	×			×	×	×	×	×				
L. minimus Gran						×	×	×	×	×		
Lichmophora abbreviata Ag.								×	×			
L. Juergensii Ag.								×				
Melosira sp.	×			×								
M. italica (Ehr.) Kütz.							×	×				
M. sulcata (Ehr.) Kütz.	×	×	×	×	×	×	×	×	×	×	×	×
Navicula sp.	×	×	×	×	×	×	×	×	×	×	×	×
N. distans W. Smith	×	×	×	×	×	×	×	×	×	×		
N. Vanhoeffeni Cleve			×	×								
Nitzschia sp.		×										
N. Closterium W. Smith	×	×	×	×	×	×	×	×	×			
N. delicatissima Cleve				×	×	×	×	×	×			
N. longissima Cleve	×					×	×	×				
N. seriata Cleve	×	×	×	×	×	×	×		×			
Pinnularia spp.						×			×			
Pleurosigma spp.	×	×	×	×	×	×	×	×	×	×	×	×
Pleurosigma angulatum (Quek.) W. Smith							×	×				

TABLE I (*Continued*)

Species	Months											
	J	F	M	A	M	J	J	A	S	O	N	D
P. balticum (Ehr.) W. Smith							×	×				
P. decorum W. Smith							×	×				
P. elongatum W. Smith							×	×				
P. fasciola (Kütz.) W. Smith							×	×				
P. Normani Ralfs	×	×	×	×	×	×	×	×	×	×	×	×
Porosira glacialis (Grun.) Jörg.			×	×	×	×	×	×				
Raphoneis amphiceros Ehr.							×					
R. surirella (Ehr.) Grun.			×			×						
Rhabdonema adriaticum Kütz.							×	×				
R. arcuatum (Lyngb. ? Ag.) Kütz.						×	×	×				
Rhizosolenia spp.	×	×	×	×	×	×	×	×	×	×	×	×
Rh. alata Brightw.	×	×	×	×	×	×	×	×	×	×	×	×
Rh. Bergonii Perag.				×								
Rh. calcar-avis M. Schultze								×				
Rh. fragilissima Berg.					×	×	×	×				
Rh. hebatata var. semispina (Hensen) Gran		×	×	×	×	×	×	×	×			
Rh. imbricata var. Shrubsolei (Cleve) Schröd.		×	×	×	×	×	×	×	×			
Rh. setigera Brightw.			×	×	×	×	×	×	×			
Rh. styliformis Brightw.			×	×	×	×	×	×				
Sceletonema costatum (Grev.) Cleve	×	×	×	×	×	×	×	×	×	×	×	×
Stephanodiscus astrea (Ehr.) Grun.			×									
Stephanopyxis turris. (Grev. & Arn.) Ralfs							×					
Streptotheca thamesis Shrubs.						×	×	×	×			
Synedra Gaillonii (Bory) Ehr.							×	×				
Tabellaria fenestrata (Lyngb.) Kütz.							×	×				
Thalassionema nitzschioides Grun.	×	×	×	×	×	×	×	×	×	×	×	×
Thalassiosira spp.	×	×	×	×	×	×	×	×	×	×	×	×
Th. baltica (Grun.) Ostf.			×									
Th. bioculata (Grun.) Ostf.			×	×	×	×			×			
Th. decipiens (Grun.) Jörg.	×	×	×	×	×	×	×	×	×	×	×	×
Th. gravida Cleve			×	×	×	×	×	×	×			
Th. hyalina (Grun.) Gran			×	×	×				×	×		
Th. Nordenskioeldi Cleve	×	×	×	×	×	×	×	×	×	×	×	×
Th. subtilis (Ostf.) Gran			×									
Thalassiothrix Frauenfeldii Grun.						×	×	×				
Th. longissima Cleve & Grun.			×	×	×	×	×	×	×	×		
Triceratium alternans Bail.	×								×			

ternaria and Aspergillus, possibly some to Monilia. These may have come from the air, but Sparrow's (1937) observations make it likely that most of them were in the water. The chief contribution from the data for 1933–1934 is as regards the numerical relationships.[4]

EARLY WINTER FLORA

The flora of the gulf as a whole is not only very scanty but most nearly uniform at this season, though the relative importance of the individual species may vary considerably

[4] For details as to the collections and method of analysis see Part I of this report, p. 150.

TABLE II

SEASONAL DISTRIBUTION OF THE DINOPHYCEAE OF THE GULF OF MAINE

Species	J	F	M	A	M	J	J	A	S	O	N	D
Amphidinium spp.					×							
A. oceanicum Lohm.									×			
Ceratium arcticum (Ehr.) Cleve	×	×	×	×	×	×	×	×				
C. bucephalum (Cleve) Cleve			×	×	×	×	×	×				
C. Fusus (Ehr.) Duj.	×	×	×	×	×	×	×	×	×	×	×	×
C. lineatum (Ehr.) Cleve	×		×	×	×	×	×	×	×	×	×	×
C. longipes (Bail.) Gran	×	×	×	×	×	×	×	×	×	×	×	×
C. macroceros (Ehr.) Cleve								×	×	×		
C. tripos (O. F. M.) Nitzsch.	×	×	×	×	×	×	×	×	×	×	×	×
C. tripos var. atlanticum Ostf.								×				
C. tripos f. subsalsum Ostf.								×				
Dinophysis spp.				×	×	×	×	×	×	×	×	×
D. acuminata Clap. & Lach.				×	×	×	×	×	×			
D. acuta Ehr.								×				
D. arctica Mereschkowsky				×	×							
D. longi-alata Gran & Braarud				×	×							
D. norvegica Clap. & Lach.				×	×	×	×	×	×			
D. Ovum Schütt.				×	×	×	×	×	×	×		
D. robusta Gran & Braarud				×	×			×	×			
D. sphaerica				×					×			
Exuviaella spp.	×								×	×	×	×
E. baltica Lohm.	×	×	×	×	×	×	×	×	×	×	×	×
E. marina Cienkowski	×	×	×		×				×	×		
E. perforata Gran								×				
Glenodinium spp.				×	×							
G. danicum Paulsen						×	×					
G. lenticulata (Bergh) Schiller				×	×		×	×	×			
Goniaulax spp.				×	×					×	×	
G. digitale (Pouchet) Kofoid							×	×				
G. orientalis Lindemann								×				
G. spinifera (Clap. & Lach.) Diesing								×	×			
G. tamarensis Lebour			×	×	×	×	×	×	×	×	×	×
G. triacantha Jörg.						×	×	×	×			
Gymnodium spp.				×	×	×						
G. Lohmanni Paulsen				×	×	×			×			
Mesoporos asymmetricus (Schiller) Lillick										×		
M. perforatus (Gran) Lillick												×
Noctiluca Miliaris Sur.					×			×				×
Oxytoxum sp.						×						
O. gracile Schiller								×				
O. reticulatum (Stein) Schütt								×				
Peridinium spp.	×	×	×	×	×	×	×	×	×	×	×	×
P. achromaticum Levander								×	×			
P. americanum Gran & Braarud				×	×				×	×		
P. breve Paulsen			×		×				×	×		
P. brevipes Paulsen				×	×	×	×	×	×			

TABLE II (Continued)

Species	Months											
	J	F	M	A	M	J	J	A	S	O	N	D
P. conicoides Paulsen				×	×	×	×	×	×	×	×	×
P. conicum (Gran) Ostf. & Schmidt				×	×	×	×	×	×			
P. conicum var. Asamushi Abé				×	×				×	×	×	×
P. crassipes Kofoid								×				
P. curvipes Ostf.				×	×							
P. denticulatum Gran & Braarud			×	×	×				×	×		
P. depressum Bail.			×	×	×	×	×	×	×	×	×	×
P. divergens Ehr.				×	×	×	×	×	×			
P. excentricum Paulsen								×				
P. Granii Ostf.	×		×		×	×	×	×	×	×		
P. globulus Stein								×				
P. globulus var. ovatum (Pouchet) Schiller				×	×	×	×	×	×			
P. globulus var. quarnerensis Br. Schröder				×	×	×	×	×	×			
P. hangoei Schiller				×	×	×						
P. minusculum Pav.	×							×	×			×
P. monacanthum Broch				×	×							
P. obtusum Karsten								×	×			
P. pallidum Ostf.								×	×	×		
P. pellucidum (Bergh) Schütt				×				×	×			
P. pentagonum Gran					×				×			
P. pyriforme Paulsen					×			×				
P. roseum Paulsen					×	×	×	×				
P. rotundum (Lebour) Schiller								×	×			
P. simplex Gran & Braarud	×	×	×	×	×	×			×	×		
P. Steinii Jörg.							×		×	×		
P. subinerme Paulsen								×	×			
P. Thorianum Paulsen					×	×						
P. triquetrum (Ehr.) Lebour			×	×	×	×	×	×	×			
P. trochoideum (Stein) Lemm.			×	×	×	×	×	×	×			
P. variegatum Peters						×						
Phalocroma parvulum (Schütt) Jörg.	×		×									
Prorocentrum micans Ehr.	×				×				×	×	×	×
P. minimum Schiller												×
P. Scutellum Schröder					×	×	×	×	×			
Protoceratium reticulatum (Clap. & Lach.) Butschli								×				
Pyrophacus horologium Stein								×				

in different parts of the gulf even then (Fig. 1).   Species dominant in the winter flora are:

Coscinodiscus excentricus	(temperate neritic)
C. centralis	(boreal oceanic)
Thalassionema nitzschioides	(boreal neritic)
Ceratium longipes	(boreal oceanic)
C. tripos	(temperate oceanic)
Prorocentrum micans [5]	(temperate neritic).

[5] P. micans has also been recorded in small numbers in spring and more abundantly at Woods Hole in summer (Fish, 1925).

TABLE III

SEASONAL DISTRIBUTION OF THE COCCOLITHOPHORIDACEAE, SILICATAE, ET AL. OF THE GULF OF MAINE

Species	Months											
	J	F	M	A	M	J	J	A	S	O	N	D
Acanthoica acanthifera Lohm						×		×				
A. acanthos Schiller	×	×	×			×			×	×	×	×
A. coronata Lohm			×									
A. monospina Schiller			×						×			
Calyptrosphaera oblonga Lohm									×			
C. uvella Schiller			×									
Coccolithus pelagicus (Wall.) Schiller	×			×	×	×	×	×	×			
Discosphaera tubifer (Murr & Blackman) Lohm			×									
Lohmanosphaera spp	×		×									
L. adriatica Schiller			×									
Pontosphaera Bigelowi Gran & Braarud								×	×			
P. Huxleyi Lohm	×		×		×	×	×	×	×	×		
P. ovalis Schiller	×											
Rhabdosphaera stylifer Lohm	×		×						×	×		
R. tubulosa Schiller	×											
Scyphosphaera Apsteini Lohm			×									
Syracosphaera spp			×						×	×		
S. mediterranea Lohm			×									
S. pulchra Lohm									×	×		
Dictyocha fibula Ehr	×		×		×	×	×	×	×			
D. fibula var. messanensis (Haek.) Lemm						×						
D. fibula f. spinosa Lemm					×							
Distephanus speculum (Ehr.) Haek	×			×	×	×	×	×	×			
Ebria antiqua Schulz												×
E. tripartita (Schum.) Lemm				×			×	×	×			
Gloeocapsa spp	×								×			
Merismopedia spp	×											
Halosphaera viridis Schmitz					×	×	×	×	×	×		
Eutreptia Lanowi			×	×	×	×	×	×				
Phaeocystis spp				×	×							
Flagellates	×	×	×	×		×			×	×	×	×

In the year 1933 the diatom genus Coscinodiscus,[6] and most commonly but not invariably *C. excentricus* was dominant in the western and northern coastal belts, forming up to 48% by number of the total (up to 1,700 cells per liter locally, near the tip of Cape Cod) accompanied by Ceratium, smaller peridinians etc. in the numerical proportions shown in Fig. 2. Coscinodiscus also dominated in the western side of the basin (Fig. 1) in about the

[6] Thirteen species of Coscinodiscus have been found in the gulf (Table I), but only 3 of them are significant items in the plankton, i.e., *C. excentricus*, *C. centralis* and *C. asteromphalus*, all of which are similar as to distribution.

same proportion, with Ceratium very scarce (less than 50 cells per liter), but with other peridinians in moderate abundance (about 300 per liter).   On the eastern side of the Gulf, however, Coscinodiscus formed not more than 10%, the dominants being species of Peridinium and of Exuviella (about 500 per liter) accompanied by neritic diatoms, chiefly *Melosira sulcata* and *Thalassionema nitzschioides* (not more than 250 per liter, Fig. 2). The one station for the month on George's Bank (eastern part) suggests dominance at the time by a more complex flora—with Ceratium, other peridinians, Coscinodiscus and various neritic diatoms in more nearly equal proportions (Fig. 2).

Essentially the same mutual relationship of the dominant members, i.e., Coscinodiscus, other neritic diatoms, Ceratium and other Peridinians, appears to have persisted at least through January of the winter in question, to judge from the close correspondence between

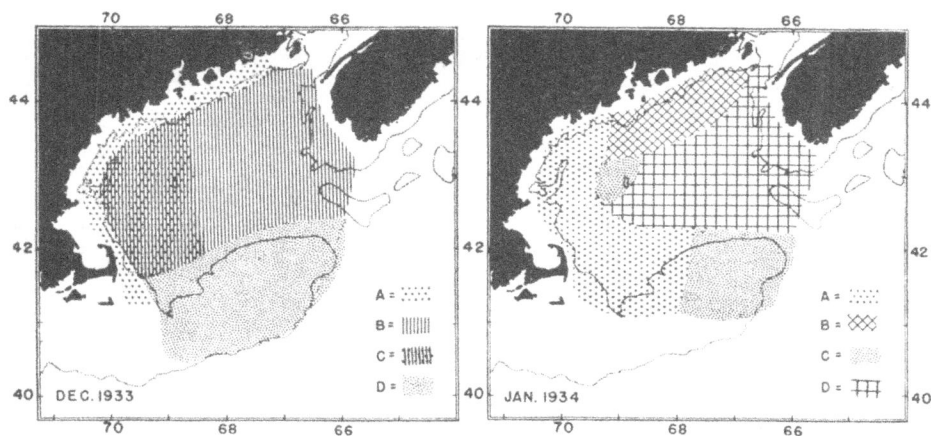

FIG. 1.   Distribution of phytoplankton communities in the Gulf of Maine.   December 1933 (left); *A*, Coscinodiscus; *B*, Peridinians; *C*, Coscinodiscus and peridinians; *D*, Neritic diatoms, plus mixture of Coscinodiscus and peridinians.   January 1934 (right); *A*, Coscinodiscus; *B*, Sceletonema and neritic diatoms; *C*, Winter neritic diatom and peridinian community; *D*, Phytoplankton nearly or completely absent.

the percentages at corresponding stations for that month with those for December (Figs. 1, 2).   Only on George's Bank may there have been any marked alteration, i.e., development of dominance by Thalassionema—nor is it certain that even this indicates any significant alteration, the January station having been located on the western part, the December station on the eastern.

It is during the mid-winter that the diatom genus Coscinodiscus attains its richest development, for although it has no definite flowering in the sense of that of Thalassiosira, Chaetoceros or even Rhizosolenia, it so increases in numbers from October on, at a time when other diatoms are disappearing from the waters, that the winter counts are frequently 50–100% Coscinodiscus.   Because of its large size the genus also frequently dominates volumetrically in samples in which it is not numerically important.   Data for earlier years suggest, however, that during the winter of 1933–1934 dominance by Coscinodiscus involved a greater proportion of the gulf than is usual, for in late December and early January

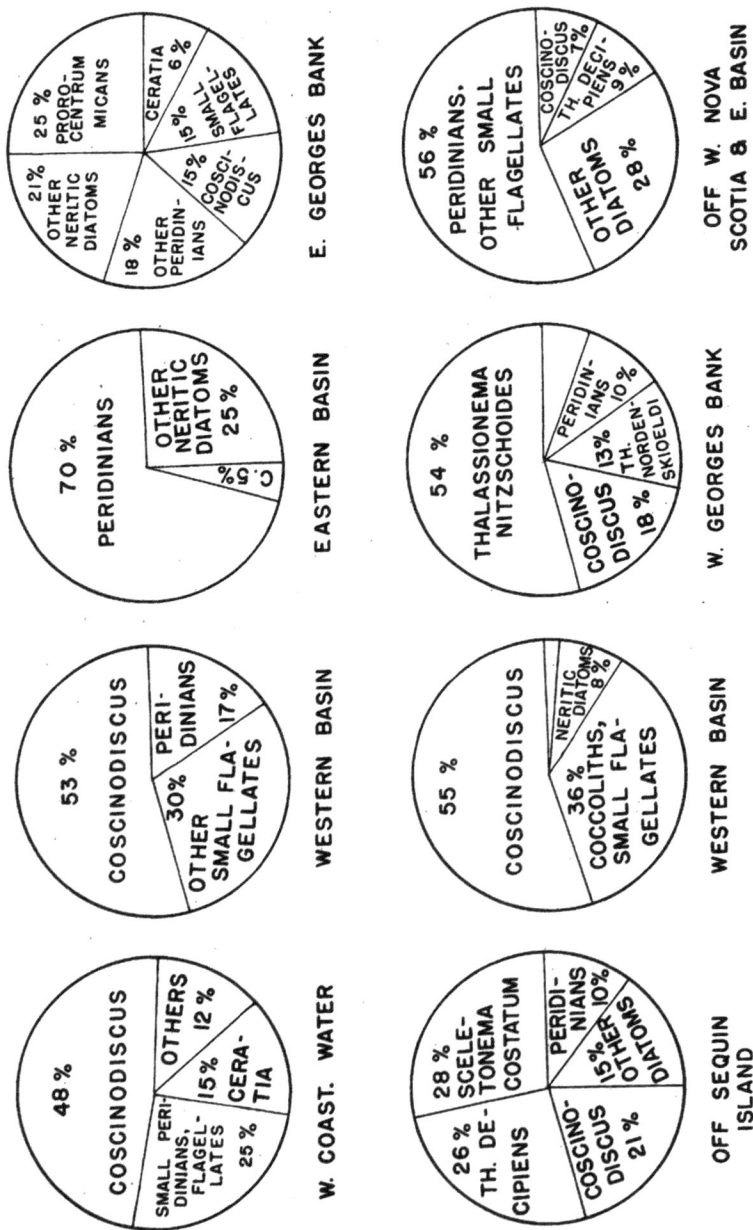

Fig. 2. Percentage composition of the phytoplankton at representative localities in the Gulf of Maine; above, December 1933 (C. = Coscinodiscus); below, January 1934.

of 1920–1921, it was dominant at one of the stations occupied by the "Halcyon" along the western and northern shores of the gulf, i.e., near the mouth of the Merrimac River, but was less important than Ceratium (though still the most numerous diatom) in the eastern side of the basin, in the Fundy deep and off western Nova Scotia (Bigelow, 1926, p. 395). Coscinodiscus was, in fact, only a minor element in the vicinity of Cape Ann in the winter of 1912–1913, *Ceratium tripos* (with occasional *C. Fusus* and peridinians) being dominant at the end of November and December with only occasional diatoms (Coscinodiscus, Rhizosolenia, Chaetoceros); and codominant with the diatoms Chaetoceros and *Thalassiothrix nitzschioides* in January (Bigelow, 1914, p. 404).

MIDWINTER MODIFICATIONS

One or other of the following species of diatoms (mainly boreal)—some of which are so scarce during the early winter that only a few chance specimens appear in collections—may show enrichment in late December or January, either in the western coastal belt or on the southern rim of the gulf:

Biddulphia aurita	(boreal neritic)
Chaetoceros atlanticus	(boreal oceanic)
Ch. borealis	(boreal oceanic)
Ch. convolutus	(boreal oceanic)
Ch. decipiens	(boreal oceanic)
Ch. densus	(temperate oceanic)
Eucampia zoodiacus	(temperate neritic)
Leptocylindricus danicus	(boreal neritic)
Melosira sulcata	(boreal tychopelagic)
Rhizosolenia alata	(temperate oceanic)
Rh. hebatata var. *semispina*	(boreal oceanic)
Sceletonema costatum	(boreal neritic)
Thalassionema nitzschioides	(boreal neritic)
Thalassiosira decipiens	(boreal neritic)
Thalassiothrix longissima	(boreal oceanic).

The magnitude of this development may be illustrated for 1934 by the accompanying table for George's Bank:

Species	Cells/liter, December, 1933	Cells/liter, January, 1934
Melosira sulcata	90	287
Thalassionema nitzschioides	265	2795
Coscinodiscus excentricus	200	200
Other neritic diatoms	65	1225
Peridinians	990	550

A mild flowering of *Rhizosolenia alata* also takes place in winter in the inner parts of the gulf, at least locally. In 1925 for example, *R. alata* dominated in Cape Cod Bay from mid-November through January and early February, and in 1934 Rhizosolenia occurred in numbers as high as 1,300 per liter (16% of the total population) off western Nova Scotia in March (Fig. 3).   Fish (1925) also found a regular winter flowering of *Rhizosolenia alata* at Woods Hole, but it is not likely that this contributes to its winter peak in the western coastal belt of the gulf, since *R. alata* has not been found on the southern banks at this season.   On the other hand a distinct maximum (up to 4000 cells per liter) was recorded there in March 1934 for *Thalassionema nitzschioides*, a neritic species which usually occurs only in small numbers in the open gulf, although it has been one of the most regularly recurrent members of the gulf plankton during most of the months of record, and there is some evidence that Thalassionema, like Rhizosolenia and Sceletonema, may have two flowering periods in the gulf—the one in early spring, the other in early autumn (p. 222).

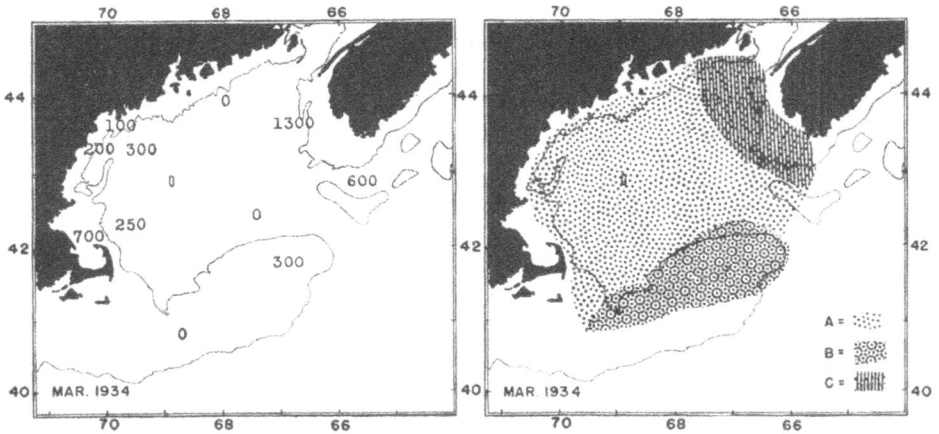

Fig. 3.   Left: *Rhizosolenia alata* in March 1934, expressed as the number of cells per liter at the level of maximum abundance, usually the surface.   Right: Distribution of phytoplankton communities in the Gulf of Maine, March 1934;—*A*, Thalassiosira; *B*, Littoral and neritic diatoms; *C*, Thalassiosira and Dinophyceae.

Sceletonema costatum also multiplied to about 500 cells per liter in the northern coastal waters of the gulf, i.e., near Seguin Island in January of 1934, a stock sufficient to make the species characteristic of the regions which it inhabits though far smaller than that which develops in European waters in early spring (Fig. 2); whereas it was either absent or only occasional on the western side of the gulf or over the southern banks at the time.   Multiplication of Chaetoceros may also take place in late January and February in some years, for although we have no evidence of this for 1934, *Ch. atlanticus*, *Ch. convolutus* and *Ch. decipiens* (combined) dominated the entire basin in late February and early March of 1920, and its western branch as late as mid-April (Bigelow, 1926, p. 421).   In 1913 Chaetoceros increased in importance relative to Ceratium from December through January, and probably in abundance as well, though counts were not made (Bigelow, 1914, p. 404).

The tychopelagic form *Actinoptychus undulatus*, which has been recorded at many stations, was codominant with Coscinodiscus and with *Biddulphia aurita* near Seal Island,

Nova Scotia, on March 23, 1920, other diatoms still being very few in number (Bigelow, 1926). The chain-forming *Navicula Vanhoeffeni*, though an arctic neritic form, was listed for April of 1932 by Gran and Braarud (1935) and was relatively abundant over the banks and in the coastal waters in March 1934, with a maximum of 4,600 cells per liter on the western part of George's Bank. A scattering of other naviculas—most commonly *N. distans*—have also been recorded at many stations in most of the months of record, in greater abundance in the Bay of Fundy than in the open gulf (Gran and Braarud, 1935); no doubt they are swept up from the bottom by the tidal currents. A diatom community of unusual composition, marked by a large variety of species of diatoms and peridinians and contrasting with the sparse flora in the surrounding waters, but similar to the flora on George's Bank at the time though not continuous with it, was encountered in the western basin near Cashes Ledge in January 1934. This same area has also been found to support exceptional associations of phytoplankton at other seasons (p. 214).

Coincident with this rise in numbers for neritic diatoms, the Dinophyceae decrease during the late winter, as shown for 1934 by the accompanying table, with the result that the net increase in the total number of vegetable cells is small (p. 155, Pt. I).

PERCENTAGE OF POPULATION

	Western Coastal Waters		George's Bank	
	December	January	December	January
Neritic diatoms............	7%	70%	21%	75%
Coscinodiscus..............	50%	20%	15%	15%
Peridinians................	20%	4%	48%	8%

The chief variations recorded from year to year in the character of the mid- and late winter flora are in the relative abundance of Coscinodiscus as compared with Ceratium and with other peridinians as just described, and in the abundance of neritic diatoms of the species enumerated above, any one of which may become dominant temporarily in one part of the gulf or another.

Spring Flora

The spring ushers in such sudden changes in the composition of the flora that it then enters upon the most interesting successional stages of its yearly cycle, even apart from the great outburst of growth (p. 156). The most obvious change from the winter state lies in the outburst of diatoms, but of species other than those characteristic of the preceding communities, for the latter (with few exceptions) are either absent from the vernal associations or of small importance in them. On the other hand peridinians continue to diminish in numbers as the spring advances, until the vernal diatoms reach their peak of development, when only occasional individuals are found. This impoverishment of peridinians cannot be attributed to their being crowded out by the rapidly dividing diatoms, since in 1920 it took place nearly simultaneously over all but the southeast corner of the gulf, at a season when the rich diatom flowerings were still restricted to rather definite centers, i.e., to the coastal waters, to the shallows of Brown's and German Banks, to the intervening channel and to the continental slope (Bigelow, 1926). Similarly in 1934 the peridinians had practically vanished from the gulf by the latter part of March, except off western and southern Nova Scotia where they still persisted in numbers not exceeding 1,100 per liter, though the diatoms had so far multiplied only in the western and northwestern coastal belt.

The spring flora is foreshadowed by the appearance of *Thalassiosira decipiens* and *Th. Nordenskioeldi* in small quantities, but widespread over the gulf in late winter (about 100–400 cells per liter in January 1934), *Th. decipiens* usually developing first, as it does in the Bay of Fundy also (Gran and Braarud, 1935), but reaching its maximum concurrently with *Th. Nordenskioeldi* (pp. 197, 207, Table I) (Fig. 4 [7]), rarely becoming a major element in the flora. It is now established that the flowering of Thalassiosira, for which *Th. Nordenskioeldi* is chiefly responsible in the gulf proper, originates just after the waters have reached their coldest, in shallow water in two chief centers, one along the narrow coastal sector between Boston Harbor [8] and Portland, the other in the coastal and bank waters south of Nova Scotia (Table IV). It is in the first of these centers that flowerings have been recorded earliest, i.e., before the end of February in 1920 and 1925, about mid-March in 1934, 1938 and 1939, but not until early April in 1913. In the Nova Scotian center this stage is not reached until two to four weeks later, or until the end of March in an early year (e.g., 1920), but not until about the middle of April in a late one, as in 1934, with early April as the average.

Thalassiosira, or that genus plus Chaetoceros, was also found in moderate abundance on George's Bank in the last half of April of 1913 and 1915. It is doubtful, however, whether it ever flowers there in such abundance as it does to the northward, for the rich flora encountered to the southwest part of the bank in late February and again in May of 1920 was dominated by *Chaetoceros socialis*, with other neritic diatoms in smaller numbers (Bigelow, 1926). Again in 1934 *Th. Nordenskioeldi* was in only moderate numbers (950 per liter), subordinate to other neritic diatoms on the southwestern part of the bank in January and less numerous in March (265 cells per liter), while the western part of the bank was dominated early in that May by the same Chaetoceros community that usually succeeds Thalassiosira in other parts of the gulf. Thus, if any active flowering of Thalassiosira took place on the bank during that particular spring, its course must have been run between the first and the last of April.

The flowering of Thalassiosira expands first along shore from the Boston-Cape Elizabeth center in both directions. In 1920 it had involved Massachusetts Bay and Cape Cod Bay by the end of the first week of April, i.e., within four weeks of its inception. It also reaches to the offing of Casco Bay at about the same time (to judge from 1920), expanding thence eastward to the Fundian region, where *Th. Nordenskioeldi* usually reaches its peak of development by the latter half of April. Such at least was the case in 1920 (Bigelow, 1926), in 1932 (Davidson, 1934; Gran and Braarud, 1935) and in 1934 (Fig. 5).

The eastern flowering also expands northward (usually) by the latter half of April to join the western flowering off the mouth of the Bay of Fundy; southward across Brown's Bank; and westward as well over the eastern side of the bowl generally, where more than

[7] At certain stations very close to shore neritic species of Chaetoceros may equal Thalassiosira in importance in early spring but only rarely farther from shore, until later in the season.

[8] The earliest flowering was north of Cape Ann in 1913 (Bigelow, 1926), but collections made in 1938 and 1939 have shown that the waters on the northern side of Massachusetts Bay likewise contribute to it.

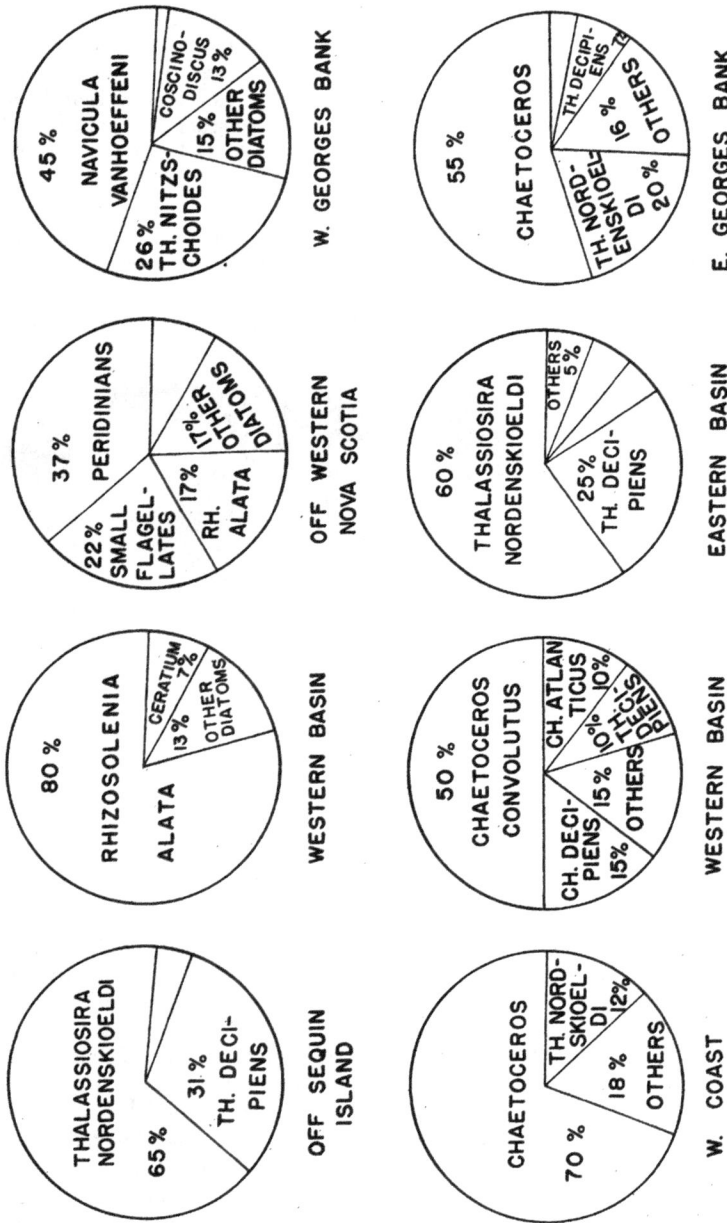

FIG. 4. Percentage composition of the phytoplankton at representative localities in the Gulf of Maine in 1934.—March (above) and late April–early May (below).

50% of the total number of cells in the richest diatom flowering yet recorded for the gulf were *Th. Nordenskioeldi* in late April of 1934, with related species accounting for an additional 30% (Fig. 4). It is probable that this is a regular yearly event in that general locality, as it is elsewhere in the gulf, for although no very rich flowering of Thalassiosira was encountered on the eastern side of the basin in 1920, whether in April or in May, it is likely that the cruises of that spring merely failed to coincide in date there with the vernal outburst.

In 1934 Thalassiosira increased in numbers in the western basin from 100 per liter in late March to between 100,000 and 200,000 in late April, though still forming little more than 10% of the total phytoplankton (Fig. 4); and it was also very prominent in the rich diatom florae encountered in one part or another of the western basin after the middle of April of 1915 and 1920, though similarly outnumbered by Chaetoceros. It may be that

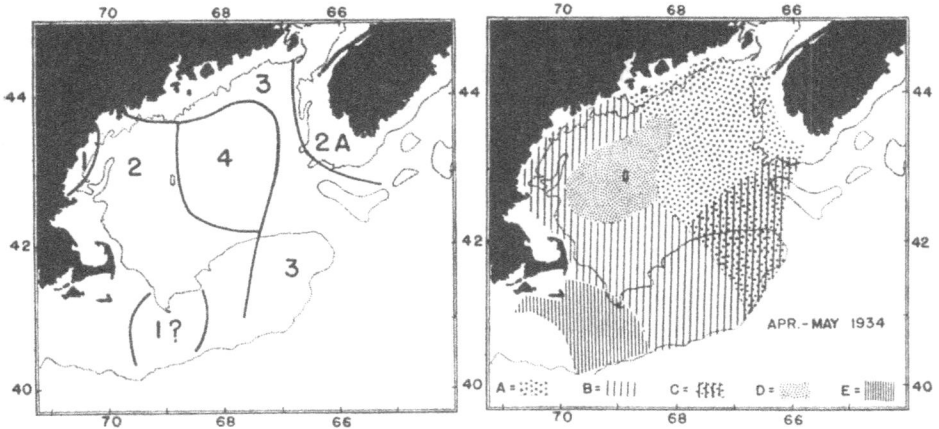

FIG. 5.   Left: Successive stages of the Thalassiosira maximum in the Gulf of Maine.   Right: Distribution of phytoplankton communities in the Gulf of Maine, late April–early May; *A*, Thalassiosira dominant, Chaetoceros minor; *B*, Chaetoceros dominant, Thalassiosira minor; *C*, Thalassiosira and Chaetoceros codominant; *D*, *Chaetoceros convolutus*; *E*, peridinians.

this mixed community develops directly as such in the western basin, but inasmuch as the vernal flowering of Chaetoceros in other parts of the gulf usually coincides with the eclipse of Thalassiosira, it is equally possible that the peak for the latter fell between the dates of the cruises in the western basin, as well as in the eastern. The fact that the flowering of Thalassiosira does not take place over all the gulf simultaneously (Table IV) makes it likely, indeed, that each survey has missed certain of these stages in some region.

The anticlockwise character of the dominant drift must condition the dispersal of Thalassiosira from its center of most active production. The flowering that originates in the western coastal belt may likely spread to Massachusetts and Cape Cod Bays and to parts of the western basin. Thalassiosira might similarly spread eastward over George's Bank from the west, though hardly from the western part of the latter to the open basins to the northeast; and the seeding of the eastern basin may be from the Nova Scotian shallows. It has already been suggested (Bigelow, 1926) that the Thalassiosira-Chaetoceros flora encountered on the eastern part of George's Bank, and to the northward toward Nova Scotia in mid-April of 1920, had its origin in the vernal Scotian drift from the east. But for Thalassiosira to spread from the Cape Ann-Cape Elizabeth sector eastward along the coast toward the Bay of Fundy would be directly against the dominant current. With small numbers of the genus present and widespread over the gulf during the months prior to its vernal outburst, it is more probable that the successive development of its peaks of abundance in different regions results chiefly from increasing activity of local reproduction, commencing as early in the season as the conditions of life become favorable for this diatom in any particular region.

COMPOSITION OF THE THALASSIOSIRA COMMUNITY

Thalassiosira Nordenskioeldi plays as dominating a rôle in the gulf as it does in the seas of northern Europe, though flowering only in early spring. Present in numbers not greater than 5,000 cells per liter in summer, and rarely more than 500 in autumn or winter, it frequently increases to more than a million per liter and to about eight billion per column in April–May. During its peak season it is regularly accompanied by *Th. decipiens*, by *Th. gravida* and by *Porosira glacialis* (p. 197), which have the same general seasonal cycle. On two occasions *Th. decipiens* was found somewhat more numerous than *Th. Nordenskioeldi*, i.e., near Seguin Island in late March 1920, and again late in April of 1934. In the latter case the codominance by *Th. decipiens* extended eastward to the ridge which separates the western basin from the eastern, where its abundance (45% of the total population and maximum of nearly 1 million cells per liter) about equalled that of *Th. Nordenskioeldi* (44%). Gran and Braarud's (1935, p. 367) statement that in 1932 decipiens "was by far the commonest of all diatoms . . . observed at all seasons . . . and nearly at every station" in the Bay of Fundy region (their observations covering March to September only), is evidence that its relative importance may increase northward, though the maximum recorded by them for that spring was only about 12,000 per liter. *Thalassiosira gravida* was recorded in abundance as great as 100,000 cells per liter in the northern part of the gulf in April by Gran and Braarud (1935), and Bigelow (1926) ranked it and not decipiens as second to Nordenskioeldi in collections made in the spring. *Th. bioculata* and *Th. hyalina* also appear in spring—are in fact limited to that season (Table I)—but neither of them attains any numerical prominence in the flora of the gulf.

Porosira glacialis,[9] a regular and important member of the Thalassiosira plankton in European waters, flowers but briefly in the Gulf of Maine, usually in late March or April, concurrent with Thalassiosira or just after the latter. It is not of importance after April, but has been found occasionally and in small numbers in May, June and August. The maximum count for it in 1934 was about 28 million cells per column and 7,050 per liter in March, but Gran and Braarud (1935) report a maximum of nearly 30,000 per liter.

Other species commonly associated with Thalassiosira flora are *Chaetoceros atlanticus*, *Ch. borealis*, *Ch. convolutus*, *Ch. debilis*, *Ch. decipiens*, *Ch. compressus*, *Ch. laciniosis* and

[9] The species for April under this name in the preliminary report (Lillick, 1938) was in reality *Thalassiosira decipiens.*

Ch. furcellatus, all of which have a seasonal distribution very similar to that of *Th. Norden-skioeldi;* also *Leptocylindricus danicus* and *L. minimus* (never as much as 1% of the spring flora).[10]   *Biddulphia aurita,* usually an insignificant item in the flora of the open gulf, may likewise be prominent locally in spring, if other diatoms be scarce, as off the Merrimac River on March 4, and off Yarmouth, Nova Scotia on April 13 of 1920 (Bigelow, 1926).   It is, however, chiefly confined to the coastal belt, there being only three records of it in the gulf outside the 100-meter contour.[11]   The Thalassiosira association also includes the arctic forms *Fragilaria oceanica, Navicula Vanhoeffeni* (limited to the early spring and chiefly to the eastern side of the gulf) and *Achnanthes taeniata,* which was recorded at 17 stations during April in 1934—also near the coast south of Nova Scotia in that same month of 1932 (Gran and Braarud, 1935), but at one station only in May.   Being unknown in the western and southern parts of the gulf, Achnanthes probably enters the latter with the cold drift from the east.

We also have record of *Eucampia zoodiacus* at a number of the April stations in 1934, both on George's and Brown's Banks and in the northern part of the gulf off the mouth of the Bay of Fundy, a distribution seemingly characteristic, for Gran and Braarud (1935) found it frequently on the northern side of the gulf in 1932 from April to June.   The accompanying table shows the relative importance of certain of these species of diatoms during the height of the Thalassiosira maximum at selected stations for late April 1934:

Species	Percent of Population					
	Off Mt. Desert Island	Off Western Nova Scotia	Off Southern Nova Scotia	Eastern Basin (3 stas.)		
Thalassiosira Nordenskioeldi......	45	54.3	39.1	59	52.7	61
Th. decipiens...........	45	23	20.8	24	27.3	31
Th. hyalina...........	2	8	0.9	4	1	3.3
Achnanthes taeniata......	2	5	1.4	4	5	0
Biddulphia aurita.......	0	0	0.2	0	0	×
Chaetoceros spp.........	5	8	27.0	13	12	3.5
Rhizosolenia imbricata var. *Shrubsolei*......	×	0.1	0.1	×	1	1
Total phytoplankton in millions per column .	1,490	13,966	285	12,817	2,122	5,312

Coincident with the multiplication of Thalassiosira in the shoal waters, the cold water peridinian *Ceratium arctica* also appears in the eastern part of the gulf, though in small numbers and sporadically.   In May 1915, for example, it ranked in abundance with *C. longipes* on German Bank, but was absent from the gulf during the rest of the year.   In

[10] The maximum count recorded for Leptocylindrus by Gran and Braarud (1935) whether along the coast of Maine or in the Bay of Fundy was 8,000 per liter.

[11] Three other species of Biddulphia are also known from the gulf, *B. alternans, B. mobilensis,* and *B. regia.*

1920 it occurred in relative abundance along the eastern side in early March; a few individuals were recorded there at the close of the diatom flowering in early May, and it reappeared along the western coast in the following mid-winter, i.e., of 1921. Intermediate forms between *C. arctica* and *C. longipes* were also numerous in these years, but Bigelow (1926) had little difficulty in differentiating between them. In the early spring of 1934 arctica-like forms again appeared in the collection along with *C. longipes*, but the numbers of both were so small and the intergradation so complete that it was impossible to distinguish between them.[12] Bigelow (1926) believes that the distribution of *C. arctica* in the gulf suggests an outside origin, and the fact that the species has never been found in even moderate abundance except on the eastern side of the gulf north of George's Bank, and that its seasonal occurrence coincides with the maximum entrance of water through the North Channel and over the Nova Scotian banks, is in line with this idea.

Duration of the Flowering of Thalassiosira. The outbursts of Thalassiosira last for four to seven weeks (average about five weeks) in the western coastal area, where, in 1913, 1920, 1921, 1934, 1938 and 1939, the peak of production for it had passed by the middle of April. Farther eastward, however, it may continue to flower in the coastal belt until well into the summer, as happened in 1915 when *Th. Nordenskioeldi* was still flourishing near Mt. Desert Island on May 15 and again on June 11. This, however, seems to be exceptional. On the Nova Scotian side of the gulf its flowering may be as brief as two weeks, as in 1920, when it was terminated by the middle of April, or it may continue for at least four weeks, as in 1932 and 1934 when *Th. Nordenskioeldi* was at its peak during the latter half of April, but had diminished almost to the vanishing point by the last week of May. The period is about the same, i.e. about 4 weeks, off the mouth of the Bay of Fundy and on the eastern side of the basin, where in 1934 it terminated before the last week in May. The growth-period for Thalassiosira is seemingly shorter, i.e. not longer than two weeks, in the waters between Cape Sable and the eastern end of George's Bank, where it had come to a close by the end of April in 1934 and in 1920 in the interval between the end of March and the middle of April, if it flowered at all in that region in that particular year. The status of the Thalassiosira community in the western basin is discussed above (p. 208).

In the open gulf Thalassiosira declines so abruptly in abundance once its peak is passed that it falls to an insignificant rank in the flora within a few days. Near Cape Ann, for example, in 1938 (Fig. 6), when samples were taken almost daily during the early spring, Thalassiosira formed about 93% of the total phytoplankton on March 30, but had entirely disappeared by April 3, similar conditions obtaining in 1939. Indications based on cruises at two to four week intervals in 1920 and 1934, combined with the data for other years, are that the genus is eclipsed with equal rapidity elsewhere (Table IV), outside the Bay of Fundy (p. 230).

Thalassiosira Nordenskioeldi is a cold water species, growing best at temperatures not higher than 5° C., the shrinkage in its numbers after its peak is passed being generally accredited to the warming of the surface layers. Its distribution in the gulf proper in the spring of 1934, as well as in the Bay of Fundy in 1932 (Gran and Braarud, 1935) is consonant with this, except where it may be subject to other influences, as in the western side of the gulf where it was no longer dominant after mid-April, and on the coastal banks south

[12] The two were combined as *C. longipes* in the preliminary report (Lillick, 1938).

of Cape Sable, whence it had disappeared by the end of May, though the surface temperatures still continued lower than 5° C.

Vernal Flowering of Chaetoceros. As the flowering of Thalassiosira declines, the diatom genus Chaetoceros tends to take its place over the gulf generally (Table IV). Appearing in the coastal waters by the end of January or early February, Chaetoceras awaits, for its rapid multiplication, the time when the surface waters warm above 5–6° C. From this date on, however, it enters upon a period of expansion as rapid and nearly as intense as the outburst of Thalassiosira which precedes it, ordinarily reaching its maximum about the time Thalassiosira falls to its minimum. Although the data for 1920 ostensibly showed Chaetoceros as at its peak at about the same date as Thalassiosira, i.e., in the early part of April, other records make it more likely that it was then in process of replacing Thalassiosira, and that the collections of 1920 had missed the earlier stages in the alternative development. In the vicinity of Gloucester, Massachusetts, in 1938 this alternation in dominance took place within the period of a few days (Fig. 6), as it did again in 1939. Nor is it probable that the interval involved exceeds a week or ten days, in any part of the open gulf.[13]

Regional expansion of the Chaetoceros flowering closely parallels that of Thalassiosira (Table IV), its development being earliest in the western coastal belt south of Cape Elizabeth, and on the western part of George's Bank; somewhat later off southern Nova Scotia. In 1938 and 1939 Chaetoceros was dominant off Gloucester by the end of March; not, however, in the western coastal belt until after the middle of April in either 1920 or 1934. In some years Chaetoceros becomes dominant equally early in the season on the western side of the deep basin, as in 1920 and 1934—probably also in 1932—though in other years (e.g., 1915) this stage in the cycle may be delayed there until early May.

Rich flowerings of Chaetoceros are equally to be expected by the middle or end of April on George's Bank. In 1913, for example, the entire northern half of the latter supported an abundant Chaetoceros flora during the latter half of the month. In 1920 *Ch. socialis* dominated on the southwestern part of the bank both in late February and again in May (the only record of this species in abundance in the gulf at any season), but with other members of the genus dominant on the southeastern part in April (Bigelow, 1926, p. 422). Again in 1934 Chaetoceros had displaced Thalassiosira over the western half of the bank by about the middle of April, and on the eastern end by the end of the month. But the irregularity of these occurrences makes it doubtful whether rich flowerings of Chaetoceros normally develop simultaneously over the bank as a whole at any time during the year. Chaetoceros dominated the phytoplankton equally early west of Nova Scotia, and between Cape Sable and the northeastern tip of George's Bank in 1920, i.e., by mid-April. In 1934, however, the coastal banks on this side of the gulf still supported a monotonous flora of Thalassiosira as late as the last week of April, though Thalassiosira and Chaetoceros were in roughly equal abundance south from Cape Sable. A Chaetoceros plankton does not usually develop in the northern coastal belt (i.e., east of Portland) until the latter part of May. In 1932, for example, Chaetoceros was about as numerous as Thalassiosira from Penobscot Bay to the Bay of Fundy by the end of that month (Gran and Braarud, 1935), while in 1915, the genus was first found dominant near Mt. Desert Island on June 11, following the prolonged flowering of Thalassiosira already noted (p. 211) for that particular year.

It is not certain whether Chaetoceros flowers actively on the eastern side of the basin

---

[13] See, however, Davidson's (1934) graphs for the Bay of Fundy; also p. 230.

every year, for though it dominated the rich plankton there during the first half of May 1915, it was not numerically important in that general region in late April of 1934, though well represented among the Thalassiosira community (Chaetoceros, about 5%, Fig. 4). If Chaetoceros flowered at all in this particular area in 1934, it must have done so in mid-May, for it was only sparsely represented there in late April (Fig. 4), and diatoms of any sort were very scarce there by the end of May.

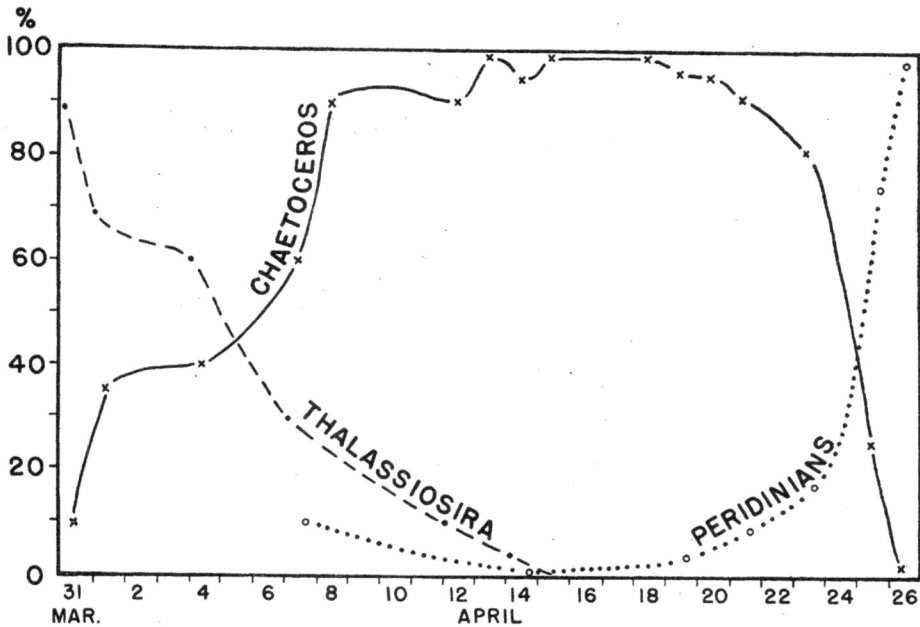

Fig. 6. Percentages of Thalassiosira, of Chaetoceros, and of Peridinians in the total phytoplankton off Gloucester, Massachusetts, March–April 1938.

At their respective peaks of abundance (Fig. 4) Chaetoceros ranks second to Thalassiosira in numerical strength, the maximum numbers of cells per column so far recorded for the two in the spring of 1934 being as follows, at localities:

Locality	Chaetoceros	Thalassiosira
Off Seguin Island	1,030,000,000	234,000,000
Mouth of Bay of Fundy	490,000,000	7,430,000,000
Eastern Basin	1,229,000,000	11,900,000,000
Eastern Basin	427,000,000	7,260,000,000
Eastern Basin	193,000,000	5,066,000,000
Southwest edge of Nova Scotian coastal bank	24,000,000	1,165,000,000
Eastern Basin	66,000,000	424,000,000
Brown's Bank	57,000,000	88,000,000
North Channel	95,000,000	131,000,000
Eastern George's Bank	53,000,000	87,000,000

Composition of the Chaetoceros Flora.   At the height of its development the Chaetoceros flora includes a greater number of species than does the Thalassiosira flora, as many as 30 members of Chaetoceros alone often occurring together, though the composition of the community is relatively uniform over the gulf as a whole.   In six years out of the ten of record *Ch. debilis* has dominated, sometimes even forming nearly a "closed stand" as it did off Gloucester in April 1939.   In other years, as in 1938, *Ch. constrictus* may outnumber *Ch. debilis*, the two making up 90–95% and 0.1–5% of the flora respectively.   *Chaetoceros compressus, Ch. (diadema) subsecundus* and *Ch. decipiens* have also been found dominant locally along the western and northern coasts in still other years, e.g., 1920, 1934, 1938 and 1939; and *Ch. densus* in 1915.   A typical Chaetoceros association for late April (as in 1934) includes the following species, most of which do not appear in the gulf until February or March:

Species	Percent of Flora
Chaetoceros affinis	2.0
Ch. atlanticus	0.5
Ch. borealis *	0.3
Ch. compressus	7.0
Ch. constrictus	3 0
Ch. convolutus	0.5
Ch. crinitis	
Ch. debilis	65.0
Ch. decipiens	1.0
Ch. densus	
Ch. didymus	2.0
Ch. furcellatus	0.3
Ch. laciniosus	2.0
Ch. radicans	2.0
Ch. scolopendra	
Ch. socialis	2.0
Ch. subsecundus	2.0
Ch. teres	1.0

* *Chaetoceros borealis* and *Ch. convolutus* so intergrade that in all probability they should finally be combined under the older name *Ch. borealis* Bailey (1854), as Gran (1904) and Føyn (1929) have already argued.

The particular species of Chaetoceros that sometimes appear at the close of winter (p. 203) (*Ch. atlanticus, Ch. convolutus, Ch. borealis, Ch. decipiens*) do not usually attain numerical importance in spring though persisting as minor elements of the flora.   This group may, however, dominate throughout the spring in one or another part of the basin in years when a pronounced peak of Thalassiosira fails to develop there, sometimes augmented by *Ch. densus, Ch. debilis* or related species.   In 1915, for example, *Ch. densus* occupied that rôle in the very abundant plankton in the western basin off Cape Ann through May.   Again in 1920 this same group of species occupied much of the northwestern part of George's Bank through March, April and early May, though with *Ch. socialis* dominant on the southwestern portion of the bank.   This group was not detected at all in January or March of 1934, but *Ch. convolutus*, accompanied by *Ch. atlanticus, Ch. decipiens* and *Ch.*

borealis (species present over the entire gulf at the time, though insignificant elsewhere in relative abundance), dominated the scanty flora in the northern part of the western basin late that April (Fig. 4). The limits of this flora were sharp, especially in the north, and it persisted in this same general region until late in May, though with its boundaries shifting a few miles inshore (cf. Fig. 5 with Fig. 7), but it had disappeared by late June. No satisfactory explanation for its temporary development is suggested.

Of the 30 odd species of Chaetoceros *Ch. decipiens* is the most persistent in small numbers throughout the year, though it never occurs in great abundance within the gulf at any season (for maximum count, see p. 221).

Duration of the Chaetoceros Flowering. The duration of the Chaetoceros flowering varies as widely as does that of Thalassiosira (Table IV). In shoal water near Cape Ann in 1938 the genus dominated the surface layers for three weeks only, i.e., until the last

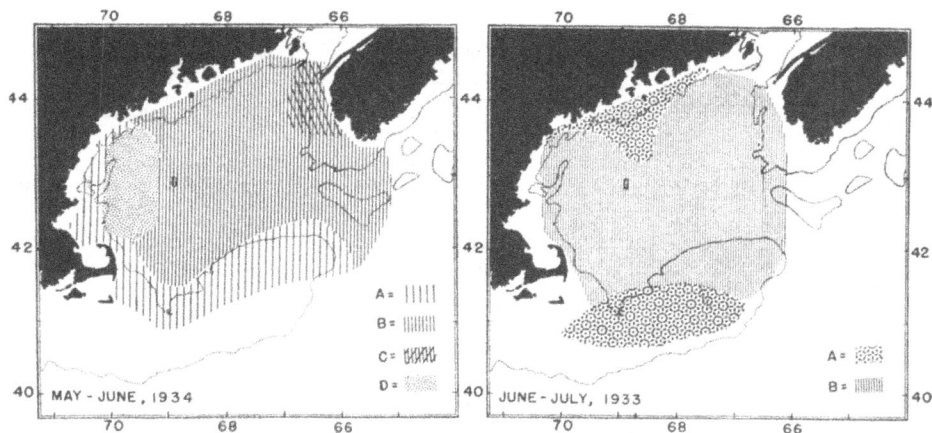

Fig. 7.   Distribution of phytoplankton communities in the Gulf of Maine.   Left: Late May–early June 1934; *A*, Chaetoceros; *B*, Peridinians; *C*, Chaetoceros and peridinians; *D*, *Chaetoceros convolutus*. Right: Late June–early July 1933; *A*, Chaetoceros and neritic diatoms; *B*, Peridinians and coccoliths.

week of April; and for about six weeks between Cape Ann and Cape Elizabeth in 1915 and 1920, to terminate by the end of May.   From Cape Ann southward to and including George's Bank Chaetoceros may flourish for perhaps four weeks, disappearing during the first days of May as in 1913, or by the end of the month as in 1915 and 1920.   In other years, however, the Chaetoceros community may persist in moderate abundance in the western coastal belt throughout the early summer, as happened in 1914 and 1934 (Fig. 7). In the shoaler waters off southern and western Nova Scotia Chaetoceros flowered in abundance until the first part of May in one year (1915), and until the middle or end of the month in others (1932, 1934), i.e., for about four weeks.   Its flowerings have been found to last from four to six weeks or even longer, in the northern coastal belt and off the mouth of the Bay of Fundy, to diminish late in May in an early year (e.g., 1932), but not until mid- or late June in a late one, as in 1915.   In the western basin the period lasts 5–6 weeks, i.e., until late May (1920, 1934) or early June (1915), and about equally long in the eastern basin to judge from the fact that the waters there were free from diatoms of any sort by the end of May or by early June in each of those years.

The fact that in 1938 the percentage of Chaetoceros fell in the course of the five day period between April 22 and 26 from 98% of the phytoplankton to less than 1% illustrates the suddenness with which its numbers may decline (Fig. 6) once its peak of abundance is passed. Sundry observations make it likely that this applies over the gulf in general. Chaetoceros persisted, however, at scattered points along the western coastal regions in

TABLE IV

SEASONAL CYCLE OF THALASSIOSIRA AND CHAETOCEROS

Area	Thalassiosira			Chaetoceros		
	Initial Flowering	Duration in Weeks	Termination	Initial Flowering	Duration in Weeks	Termination
Boston Harbor and Cape Elizabeth	Mid-March	4–7	Early to mid-April	Mid-April	ca. 3–6	Mid–late May
Massachusetts and Cape Cod Bays	Early April	2–4	Mid-late April	Mid-end of April	3–4	First–last of May
Western George's Bank	ca. 20th of March	2–3	Mid-April	Mid-April	3–4	Mid-May
West Basin	End of March	ca. 2	ca. mid-April	ca. mid-April or early May	ca. 5	Late May or June
Nova Scotia Shoal waters	Early April	2–3	ca. early May	Early May	ca. 4	Late May
Eastern George's Bank	Early April	2–3	Late April	Late April	ca. 4	End of May
Cape Elizabeth across to the Bay of Fundy	ca. April 10th	2–6 or in-definitely	ca. late May	Late May	4–6	Mid–end of June or August
East Basin	Mid-April	3–4	ca. mid-May?	Mid-May	—	—

scant numbers throughout the summers of 1913, 1914 and 1915. *Chaetoceros (criophilum) convolutus* was in fact relatively abundant off Penobscot Bay on August 14, 1914.

Phaeocystis. In 1920 Bigelow found the unicellular alga Phaeocystis practically monopolizing the surface waters in the southern part of Massachusetts Bay and in Cape Cod Bay during the latter half of April when it rivalled in its abundance the preceding peak for diatoms; it was also scattered sparsely over the western basin at the time. But this flowering was so brief that it was found neither three weeks previously nor two weeks later. In as much as we have no other record of Phaeocystis it is an open question whether its occurrence in 1920 was exceptional, or whether it flowers annually in the gulf, but so briefly and so sporadically that the other surveys have missed it.

Halosphaera. The spring is also the season of maximum frequency for the unicellular green alga *Halosphaera viridis*, which was recorded—always in small numbers—at many stations in March of 1915, and in that and the two succeeding months of 1920. It was not detected at all in 1933–1934; hence the reader is referred to Bigelow (1926, p. 459) for a discussion of its status in the gulf.

SUMMER FLORAE

Early Summer Flora. On George's Bank diatoms may continue to dominate the flora throughout the summer, though sporadically both as to dominant species and as to numerical distribution. In 1934 the bank as a whole supported a Chaetoceros community in May and June (Fig. 7), but in other summers the diatom florae have been limited on the banks to isolated areas, with sparse peridinian florae elsewhere. In 1915 and 1920, for example, communities of Chaetoceros were encountered on the southwest corner of the bank, whereas in 1933 diatoms, chiefly Chaetoceros accompanied by Thalassiosira, *Guinardia flaccida* and others, as well as by the coccolith *Pontosphaera Huxleyi*, occupied the western half (Fig. 7). *Guinardia flaccida* flowered on the western end of the bank in early July of 1913, with lesser numbers of *Eucampia zoodiacus* (Bigelow, 1926, Fig. 123), the latter a spring and summer [14] species chiefly limited to the shoaler parts of the gulf. Late in July of 1916 (Bigelow, 1926, Fig. 124) this same general region was dominated by Rhizosolenia and by the oceanic species *Thalassiothrix longissima*, a species which is probably an immigrant in the gulf, though there are scattered records of its presence inshore as well as offshore between March and October (Fritz, 1921; Bigelow, 1926; Gran and Braarud, 1935). Guinardia was flowering in July, 1914 on the northeast part of the bank, where in the same month of 1933, *Pontosphaera Huxleyi* far outnumbered the moderately abundant diatoms, chief among which were *Detonula confervacea* and *G. flaccida* (Braarud, 1934). The fairly rich phytoplankton in this same area recorded in late June and early July of 1934 was chiefly *Thalassiosira Nordenskioeldi*, *Thalassionema nitzschioides*, *Melosira sulcata*, *Nitzschia longissima*, *Peridinium conicum*, *P. Granii* and *P. simplex*, with small flagellates. Evidently, then, the diatom florae on the banks may be dominated in summer by any one of a number of species, most often perhaps by *Guinardia flaccida*.

In some years (e.g., 1912, 1914, 1915, 1933) a stock of diatoms sufficient to dominate the diminished flora—but in this case chiefly Chaetoceros—accompanied by Peridinians (especially Ceratium) in smaller numbers—also persists in the western and northern coastal belt throughout the summer months; especially is this likely to be the case in the easternmost sector of this belt near Mt. Desert Island and in the Bay of Fundy.

Typically this flora includes the following species:

Chaetoceros atlanticus	(boreal oceanic)
Ch. compressus	(boreal neritic)
Ch. cinctus	(temperate neritic)
Ch. constrictus	(temperate neritic)
Ch. debilis	(boreal neritic)

[14] Gran and Braarud's (1935) new species, *E. recta*, was not detected in 1933–1934.

Ch. decipiens	(boreal oceanic)
Ch. didymus	(temperate neritic)
Ch. furcellatus	(arctic neritic)
Coscinodiscus centralis	(boreal oceanic)
Guinardia flaccida	(temperate neritic)
Leptocylindrus danicus	(boreal neritic)
Rhizosolenia fragilissima	(boreal neritic)
Rh. hebatata var. *semispina*	(boreal oceanic)
Rh. imbricata var. *Shrubsolei*	(temperate oceanic)
Thalassionema nitzschioides	(boreal neritic)
Thalassiosira decipiens	(boreal neritic)
Th. Nordenskioeldi	(boreal neritic)
Thalassiothrix longissima	(boreal oceanic)

In other years, when the vernal warming of the surface waters is accompanied by an eclipse of the Chaetoceros flora (as described above), leaving the gulf as a whole nearly barren of diatoms, the latter are succeeded by a sparse community of peridinians, a situation developing earliest near Cape Ann and southward to Massachusetts Bay, but extending northward throughout the coastal area within a few days.   In 1938, for example, peridinians which were only 4% of the flora near the Cape on April 19th soon dominated so strongly over the few remaining diatoms that they constituted 99% by the 26th of the month (Fig. 6), and in 1915 they dominated in Massachusetts Bay by the first week of May.   Dinophycean dominance was similarly established south of Mt. Desert Island by the end of April in 1932, throughout the entire western and northwestern coastal waters by mid-June in 1934 and by early July 1933 or earlier (Fig. 7).

In the offshore basins also the waters were practically clear of diatoms by the end of May in 1934, their place having been taken by a very scant flora of peridinians, and although in 1915 this development did not involve the basin as a whole until late in June, this was found to.be a tardy year in other respects as well.

The situation is seemingly more complex in the eastern coastal belt.   Here in 1915 the diatoms had practically disappeared, and peridinians had succeeded them by early May, but they reappeared then in small numbers among the more numerous peridinians; in the summer likewise over Brown's and German Bank and in the North Channel (Bigelow, 1917; 1926).   On the other hand it seems likely that the weak growth of Chaetoceros which mingled with the peridinian population west of Nova Scotia at the end of May in 1934 (Fig. 7) was evidence of a stage in the disappearance of the diatoms, rather than of a tendency for these to persist in abundance into the summer, since peridinians had by then monopolized the waters southward from Cape Sable.

The Dinophycean flora of late spring and early summer is dominated by *Ceratium longipes*, *C. tripos* and *C. bucephalum*, with considerable numbers also of *C. Fusus* and *C. lineatum*.   Often the entire flora consists of little but this genus; however, Ceratium is usually accompanied by *Peridinium depressum*, *P. conicum*, *P. crassipes* and other species of that genus, as well as by Dinophysis and Exuviella.   The latter reaches its maximum abundance in May and June (up to about 1,000 cells per liter, Gran and Braarud, 1935), *E. baltica* and *E. marina* being the most important species.   Mixed with the peridinians are also a scattering of temperate species of diatoms, among them various species of Chae-toceros, *Rhizosolenia hebatata* var. *semispina*, *Rh. fragilissima*, *Rh. imbricata* var. *Shrubsolei* and *Thalassiothrix longissima* (Fig. 8). Water bottle samples for the summers of 1932, 1933 and 1934 [15] have also shown that Coccolithophoridaceae, especially *Pontosphaera Huxleyi*, may also be numerically important in the flora at this season. Offshore and in-shore stations are in agreement in all this, the chief regional differences in this respect being that tychopelagic diatoms appear in greater numbers at the former; whereas Coccolitho-phorids frequently dominate the summer flora offshore (Fig. 8), which rarely, if ever, happens inshore in so far as our knowledge of this area goes.

The development of summer dominance by peridinians results in part from the im-poverishment of diatoms, but in part also from multiplication by the peridinian species that are present during the vernal outburst of diatoms, combined with the appearance of additional species as well, the cyclic development in the gulf of the several species of Cera-tium being especially interesting in this connection. Of the seven species and numerous varieties of that genus which occur there, *C. tripos*, *C. longipes*, and *C. Fusus* are the most abundant; the latter is rather generally distributed throughout the year; the other two, however, show decided mutual fluctuations which have already been discussed in detail by Bigelow (1926). Briefly the situation is as follows: both species are present throughout the year, but *C. tripos* dominates the offshore plankton during the winter, while *C. longipes* is very scarce. Both species are at their minimum during the spring flowerings of diatoms —as are all other peridinians. *Ceratium tripos* also continues scarce after the diatoms diminish in numbers, but *C. longipes* then multiples to its maximum in July, when the gulf as a whole (but with the exceptions noted above) supports the most abundant Ceratium plankton of the year. During August *C. longipes* declines in amount, but *C. tripos* multi-plies, with the result that within two or three weeks the relative importance of the two species has been reversed. This reversal from dominance by *C. longipes* to dominance by *C. tripos* takes place earliest over the southern and western half of the basin, following northward and eastward, until it finally involves the northern coastal belt from Casco Bay to the Bay of Fundy by September or early October (Bigelow, 1926, Fig. 110). Fur-ther evidence that this sequence is fairly constant, although with local exceptions, is pro-vided by the data for more recent years; thus, while both of these species were present in small numbers only (though widespread) in December, January and March of 1933–1934, *C. longipes* greatly outnumbered *C. tripos* by June in most parts of the gulf. A similar situation also developed within the space of a few days near Cape Ann in late April of both 1938 and 1939. By August of 1938, however, *C. tripos* had again become the dominant peridinian in the western side of the gulf (the only region examined in that summer); whereas by September (of 1933) it and *C. longipes* were both present in roughly equal numbers.

The mutual relationship of these two species seems essentially the same on George's Bank also, for *C. longipes* was the more abundant member of the pair there in late May, but *C. tripos* definitely so in June of 1934, in July of 1933 and in August of 1932, but with the two about equal in September (1933). Both were so scarce (20–40 per liter) during the late autumn, winter and early spring of 1933–1934 that sometimes the one was recorded

---

[15] Records for previous summers were based on catches with tow nets which do not sample vegetable cells as small as Pontosphaera.

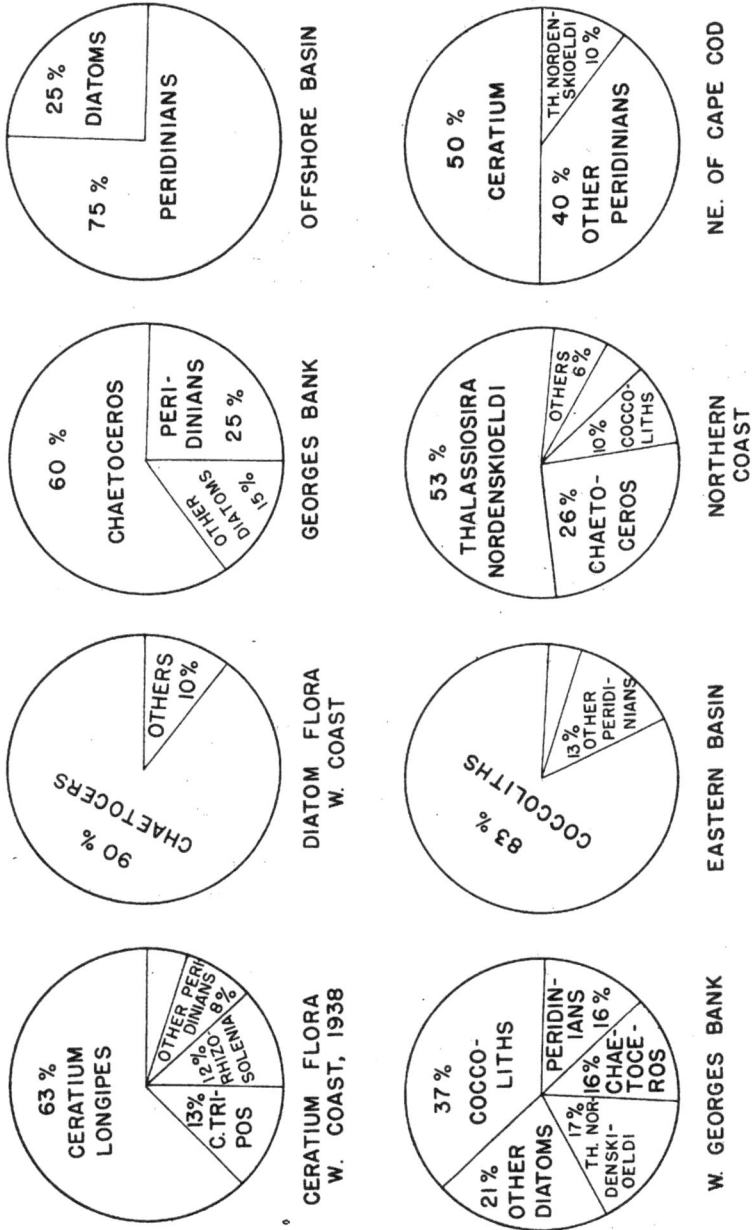

FIG. 8.   Percentage composition of the phytoplankton at representative localities in the Gulf of Maine, late May–early June 1934 (above), and late June–early July 1933 and 1934 (below).

(in October, December, January, *C. tripos*), sometimes the other (*C. longipes* in March), and sometimes neither (April, 1934). Gran and Braarud (1935) who found *C. tripos* less abundant than Bigelow's (1926) records led them to anticipate, but *C. bucephalum* more so, suggested (on the basis of one of Bigelow's (1926) microphotographs) that he might have confused the two species, but the evidence for 1933–1934 and for 1936 tends to substantiate Bigelow's conclusion that *C. tripos* is usually much more abundant than *C. bucephalum*, for when the latter was detected at all, it was always recorded as "scarce." It is often difficult, however, to distinguish between these two members of this proverbially variable genus.

LATE SUMMER FLORAE

The chief alteration in the flora during the late summer or early autumn is in connection with the second diatom flowerings, for which Chaetoceros, Rhizosolenia or Sceletonema are chiefly responsible in one part of the gulf or another. On the northern coastal belt, as exemplified by Penobscot and Frenchman's Bays near Mt. Desert Island, Chaetoceros predominates the more often than Rhizosolenia in the late flowerings, with *Ch. debilis*, *Ch. decipiens*, *Ch. compressus*, and *Ch. laciniosis* as the most abundant species (Burkholder, 1933). Gran and Braarud (1935) also found Chaetoceros in numbers as high as 300,000 cells per liter near Mt. Desert Island on August 14, 1932 (including 3,600 of *Ch. decipiens*, the highest count yet recorded for that species in the gulf), and somewhat less abundant at neighboring stations after a period of diatom scarcity. The representative species of Rhizosolenia that develop in greatest abundance inshore in late summer are *Rh. setigera*, *Rh. imbricata* var. *Shrubsolei*, and *Rh. hebatata* var. *semispina*. It is chiefly this genus which is responsible for the second flowering that occurs farther outside along the northern coastal belt and over George's Bank (Fig. 9). The late summer is, in fact, the only season when

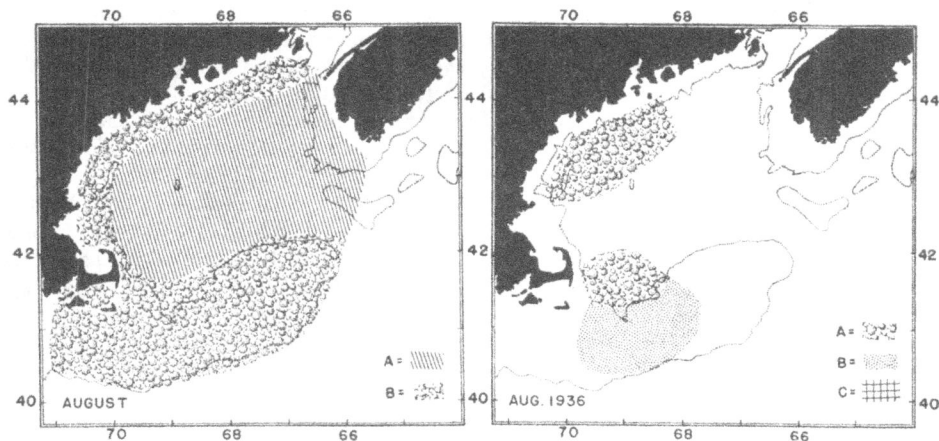

Fig. 9. Distribution of phytoplankton communities in the Gulf of Maine. Left: August, *A*, Ceratium plankton; *B*, Extent of diatom flowering. Right: August 1936, *A*, *Rhizosolenia alata*; *B*, *Rh. styliformis*; *C*, *Rh. calcar-avis* and *Sceletonema costatum*.

this genus multiplies to great abundance (see, however, p. 203), though it occurs widespread in the gulf in small numbers throughout the year.   The leading members of the late summer florae for these areas are *Rh. alata*, which usually rank first, *Rh. styliformis*, *Rh. imbricata* var. *Shrubsolei*, *Rh. setigera* and *Rh. hebatata* var. *semispina*.   *Rh. alata* is on the whole the most important member of the genus in the gulf, where, by present indications it has two flowerings, one of low intensity at the close of the winter (p. 204), and a second in August of great intensity but irregular in distribution.   Instances are: its dominance in the moderately abundant plankton of the outer Massachusetts Bay in August, 1922; off Mt. Desert Island in July, 1915; and its presence in abundance in the deep water a few miles to the north of the bank on July 8, 1913; contrasted with its presence in small numbers only on George's Bank in July of 1914 and 1916 (Bigelow, 1926, p. 447).   It was also dominant on the eastern part of George's Bank (maximum frequency, 50,000 cells per liter), the most abundant diatom (though subordinate to Ceratium) in the neighboring basin, and common in the northern part of the gulf and in the Bay of Fundy (up to 6,000 per liter) in August of 1932 (Gran, 1933; Gran and Braarud, 1935).   Although it was very scarce at two stations on the eastern side of the gulf in July 1933 (Braarud, 1934), it was again so abundant at the six stations on the western side of the gulf in August of 1936 that it colored the water (Fig. 9 —no counts of cells available).   However, it has never been found in any number during the summer in the waters off western Nova Scotia.   These records are enough to show that *Rh. alata* may be expected to flower yearly in the northern and western coastal belts, on George's Bank and to some extent over the deep basins in August (or even in July).   It is, in fact, at this season that it reaches its greatest development in these parts of the gulf.

Rh. hebatata var. *semispina* was the species dominant south of Marthas Vineyard in August (of 1914), where it formed the bulk of the catch (Bigelow, 1926, p. 443, Fig. 125). It was, however, scarce or absent from these localities in other summers, as in 1914, 1916 (Bigelow, 1926) and 1933.   Although rather generally distributed farther north in the gulf in July of that year or in September, it was nowhere dominant there, nor did Gran and Braarud (1935) find it in any numbers in August of 1932, whether in the open gulf or in the Bay of Fundy, though it occurred there regularly from April (maximum, 300 cells per liter) to September and was fairly abundant near Cape Ann in late April 1938.   Late summer or early autumn flowerings of this species thus appear to be limited to the southern side of the gulf, and are sporadic even there.

Rhizosolenia imbricata var. *Shrubsolei*, present in the Gulf of Maine from February through October and even as late as December in the Bay of Fundy, was similarly found in moderate numbers (ca. 200 per liter) over George's Bank in July of 1933, also in great abundance in Nantucket Sound in 1915, and in late August of 1936—when it was only slightly less numerous there than was *Rh. calcar-avis*, a species the northward extension of which appears to be limited by Cape Cod.   *Rhizosolenia styliformis* is far more important on George's Bank and Nantucket Shoals, where it has been repeatedly recorded in large numbers during July and August (Fig. 76), than it is in the inner parts of the gulf, as Bigelow (1926) has pointed out.   This species is typically oceanic; Bigelow accounts for its distribution in the gulf as being conditioned by the entrance of oceanic water.

Guinardia flaccida may likewise be an important component of the late summer flowering, though generally less so than Rhizosolenia, occurring characteristically in very localized but dense aggregations, instances of which have been recorded at various points close to shore in the western coastal belt as well as on the southern banks.   Thus Bigelow reported a dense flowering of it over the western part of George's Bank in July 1914, and Gran (1933) and Braarud (1934) found it an important member of the July and August flora on the eastern part of the bank in 1932 and 1933.   It also appeared, though usually subordinate to Rhizosolenia, in every region investigated in August of 1933 and 1936; and in September (of 1933) when 100,000 cells per liter were recorded near Seguin Island; Gran and Braarud (1935) likewise recorded it from Passamoquoddy Bay in September, but in small numbers only (maximum, 460 cells per liter).   This regularity of occurrence in late summer and early autumn, contrasted with the record of it at one station only in January and in June, shows that Guinardia is distinctly a warm water form in the gulf.

Sceletonema costatum, a species of major importance in the early spring diatom maximum in the waters of northern Europe, develops in considerable abundance in the Gulf of Maine or in the Bay of Fundy only in late summer, when it may multiply to 700,000 cells per liter or more.   Even then, however, there is no record of it in the gulf in such numbers as in European seas, where on occasion, there may be several million of its cells per liter. Its flowerings in the gulf are closely confined to the vicinity of the coast.   Thus, Gran and Braarud (1935) found it far more abundant (ca. 500,000 cells per liter) close to shore near Mt. Desert Island in August of 1932 than was Rhizosolenia (ca. 1,000 cells per liter); also in numbers as high as 39,000 cells per liter along the western Nova Scotian coast.   The same species was abundant on the western part of George's Bank on July 9, 1913, dominant in Nantucket Sound in August of 1936, and responsible for the second flowering of diatoms that developed in Massachusetts Bay in September of 1915 (Bigelow, 1926, p. 394), but for which no counts are available.   On the other hand Gran (1933) found only a few scattered chains of it off shore either on George's Bank or in the basin to the northward in August of 1932, and it was not detected at all in the summer of 1916 (Bigelow, 1926), showing that its aestival flowerings are not as regularly annual or as widespread, even within the zone of its most characteristic distribution, as are those of the vernal diatoms.   Sceletonema is made further interesting by the probability that it differs from most others in having two flowerings a year in the gulf, the one in late summer and early autumn, the other—far less productive—in winter (p. 204).

Gran and Braarud's (1935) record within the Bay of Fundy of 100,000 cells per liter of *Thalassionema nitzschioides* suggests that that species may also flower twice yearly here but most intensely in summer, since the only large count for it in the open gulf was for March (p. 204).   *Nitzschia seriata*, present from late January to October, also has its maximum in July or August.   It is most plentiful in the peripheral belt of the gulf, but never sufficiently so to come within the "dominant" category, the largest recorded count for it in the gulf being about 8,000 cells per liter.   *N. Closterium*, with a similar seasonal distribution, but usually very scarce, was found abundant on one occasion in August, 1936 growing on the walls of Rhizosolenia.[16]   *Thalassiothrix longissima* also calls for mention in the summer-autumn group, because it dominated a rich plankton on western George's Bank in late July of 1916, but we have few other records of its occurrence in the gulf.

Finally, to emphasize the sporadic nature of the late summer flowerings of diatoms as contrasted with the vernal, we may instance the case of *Asterionella japonica*.   In August of

[16] For the seasonal occurrence of the other species of Nitzschia in the gulf, *N. delicatissima* and *N. longissima*, see Table I, p. 196.

1912—a summer when diatoms as a whole were more than usually abundant—this species was widely distributed in small numbers along shore from Cape Elizabeth to western Nova Scotia, and in such abundance between Grand Manan Channel and Sequin Island that "the appearance of the water was noticeably soupy" (Bigelow, 1914, p. 133), with "such a sudden transition to clear water with very little plankton" near Seguin Island "that the change was plainly visible from the deck of the *Grampus*" (Bigelow, 1926, p. 431). Subsequent records of Asterionella in the open gulf are confined to small numbers in the offing of Mt. Desert Island in August, 1930 (Burkholder, 1933) and to the northern part (chiefly inshore) in August of 1932 (Gran and Braarud, 1935). It is more common in the Fundian region occurring in small numbers near St. Andrews between April and December, and more numerously in the open bay, with a distribution center near Grand Manan Island where (in 1932) it had a distinct maximum in August, though always subordinate to other diatoms (Gran and Braarud, 1935). This distribution suggests that the presence of Asterionella in such unusual abundance in the northern gulf in 1912 may have been due to a greater Fundian influence in that summer than is usual, and this hypothesis is supported by the fact that the drift of cold surface water from the bay was also more strongly developed along the northern shore of the gulf in that summer than in those of 1915, 1923 or 1925 (cf. Bigelow, 1927, Fig. 47).

Other diatom species that are present in small quantity in late summer and some of which are limited to that season, are listed in Table I.

During the second diatom flowering the peridinians as a rule play a very insignificant rôle in the flora because of the great numbers of diatoms, though the same peridinian species which are present earlier in the summer persist in undiminished quantity and with additional species as the summer progresses (Table II).

The bowl of the gulf outside the 100-meter line is only slightly affected by the second flowering of diatoms, and the chief alteration in the percentual composition of the phytoplankton which takes place there from early to mid- and late summer is that *Pontosphaera Huxleyi* (the only coccolith which is of numerical importance in the gulf) may so increase in abundance as to raise its numbers far above those of the peridinians locally, as was the case at two stations on the northwestern part of the gulf in August, 1932 (Gran and Braarud, 1935), on the eastern part of George's Bank, July 16–18, 1933 (Braarud, 1934), and in the southeastern part of the basin on that same cruise,[17] as well as in the preceding August (Gran, 1933, Table I). Coccoliths, in fact, are at their maximum abundance at that season. Such extreme development of Pontosphaera appears, however, to be restricted to rather definitely circumscribed centers, for other mid- and late summer stations have yielded larger numbers of peridinians than of coccoliths.[18]

Apart from Pontosphaera, peridinians continue to dominate the bowl of the gulf throughout the late summer and early autumn, Ceratium being most conspicuous in all regions, accompanied by lesser numbers of Peridinium and of the smaller Dinophyceae and only occasional diatoms (Fig. 10), except as these may extend out into the deep basins from

[17] The latter was the maximum count so far recorded for Pontosphaera in the gulf, 308,000 per liter at the level of greatest abundance (Braarud, 1934, Table 1).

[18] The methods used to concentrate the material of 1933–1934 were such as to distort most members of this group beyond specific recognition. For a further account of these forms, see Gran and Braarud (1935).

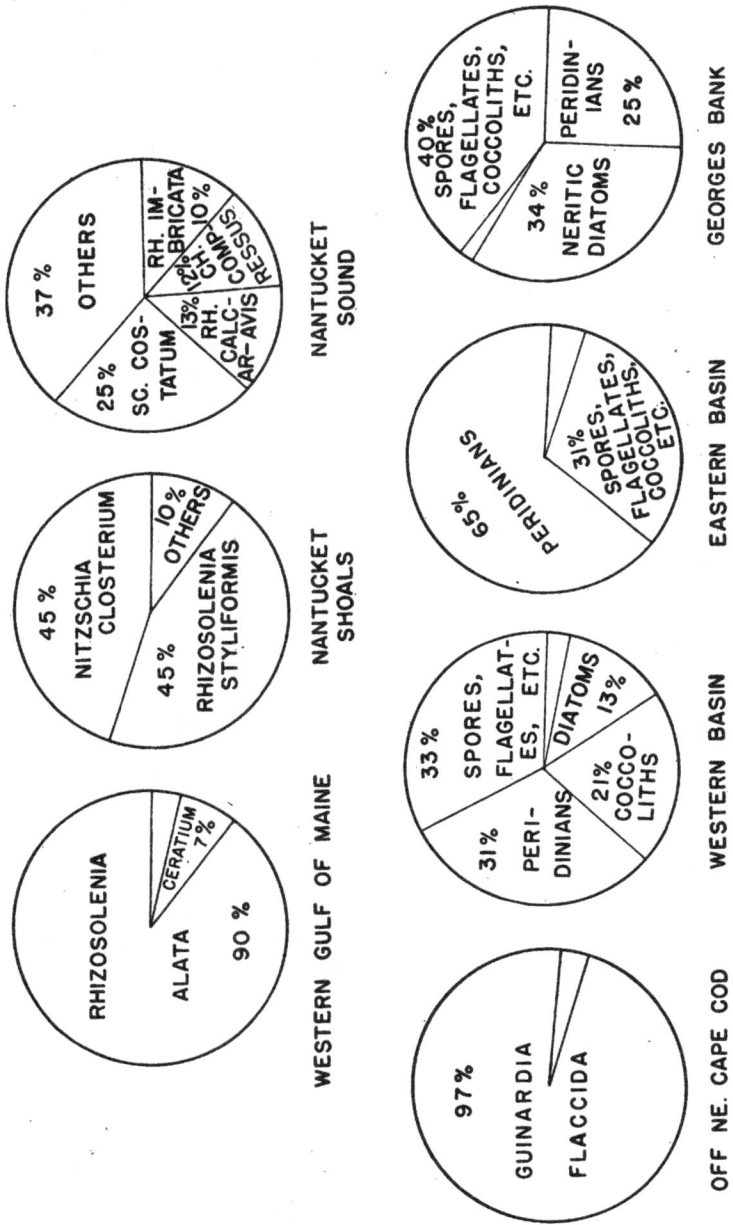

Fig. 10. Percentage composition of the phytoplankton at representive localities in the Gulf of Maine, August 1936 (above) and September 1933 (below).

the western coastal belt and from George's Bank.   The Rhizosolenia flowering of August 1936, for example, spread well out into the western part of the basin (Fig. 9), where *Rhizosolenia alata* formed more than 90% of the population, with other species of that genus, *Guinardia flaccida* and Chaetoceros present in small numbers, in combination with the assemblage of *Ceratium tripos, C. Fusus, C. longipes* and Peridinium that is usual for the season.

August–September is the season of maximal abundance for Dinophysis which is found regularly through late spring, summer and autumn, but at all times in small numbers.   Of the 7 or 8 species recorded *D. acuminata* and *D. norvegica* are the most prominant, Gran and Braarud (1935) having recorded up to 1,100 per liter of the former and 1,560 of the latter. The largest counts for Goniaulax [19] in the open gulf were for July and for August, respectively 2,800 and 1,040 per liter; Gran and Braarud, 1935.

The silicoflagellate *Distephanus speculum* has also been found somewhat more common in August and September (up to 500–600 per liter) than at other times of year.

It is in July or August, if ever, that tropical species of planktonic plants may be expected to drift northward from outside into the Gulf of Main (Bigelow, 1926, p. 393). But the only cases of this sort so far recorded—apart from the fragments of Sargassum that are sometimes reported or picked up on the offshore banks, or even on German Bank (Bigelow, 1917)—have been occasional specimens of *Ceratium macroceros* detected among the boreal species of that genus off the Merrimac River in December 1920, and in small numbers at all the offshore stations in August 1936 (Fig. 9; Fig. 4, Part I of this report).

AUTUMNAL FLORA

In the coastal belt between Mt. Desert Island and Cape Cod on the one hand, and on George's Bank on the other, a mixed flora of neritic diatoms and of peridinians may persist for an indefinite period after the climax of the late summer flowering of diatoms (Fig. 11). *Sceletonema costatum*, for example, has been found in fair abundance in shoal water near Mt. Desert Island in September (1922, 1931, 1932 and 1933), also in Massachusetts Bay, where it formed nearly 100% of the moderate diatom catch late in that month of 1915. *Guinardia flaccida* was widespread in small numbers (not over ca. 150 cells per liter) over the offshore basins in September of 1933, and was flowering richly both in the northwestern part of the gulf near Seguin Island and on the eastern part of George's Bank (100,000 and 23,000 per liter), where it was, in fact, at its peak of abundance for the year.   *Thalassiosira gravida* and *Coscinodiscus concinnus* accompanied by species of Chaetoceros, Rhizosolenia, Sceletonema, with a large variety of neritic and tychopelagic diatoms in small numbers, were similarly responsible for a moderately abundant flora off Penobscot Bay in mid-September of 1915. *Chaetoceros decipiens, Rh. setigera, Th. decipiens* and *Thalassiothrix longissima* played this same rôle near Mt. Desert Island in early October of that same year, and there was evidence of a brief flowering of *Thalassiothrix longissima* at various points along the western shore of the gulf at the time.   Other neritic and tychopelagic diatoms also have been known to grow in relative profusion (over 1,000 cells per liter) locally in shoal water in early autumn; *Melosira sulcata*, for example, off Cape Sable and *Thalassionema nitzschioides* off Cape Cod during September 1933.

Such variety in the species which have been found flowering in moderate abundance in autumn in so few years of record added to the list of other species that occur with them in

[19] *G. tamarensis* is the only species of this genus that has been found with any frequency in the gulf.

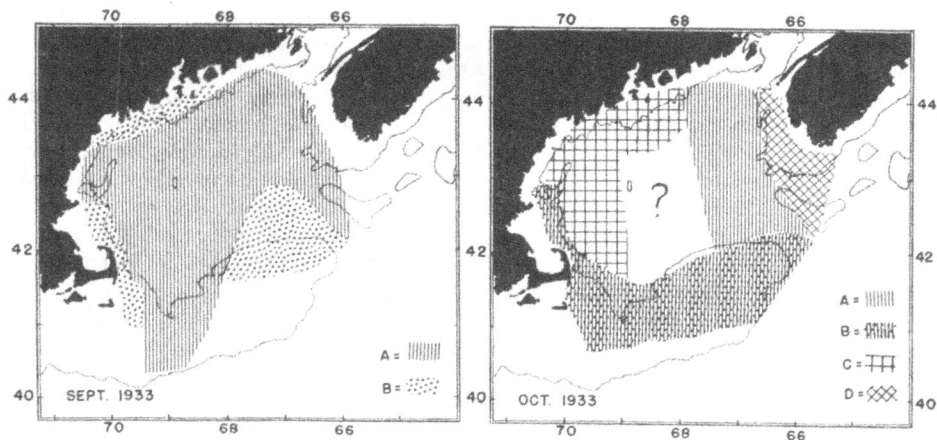

FIG. 11.   Distribution of phytoplankton communities in the Gulf of Maine.   Left: Early September 1933, *A*, Dinophyceae; *B*, Dinophyceae and diatoms.   Right: Late October 1933, *A*, Dinophyceae; *B*, Dinophyceae and Coscinodiscus; *C*, Phytoplankton nearly or entirely absent; *D*, *melosira sulcata* and littoral diatoms.

small amounts (see following Table), emphasizes the complexity of the diatom flora of the shoaler parts of the gulf at this season, as does the fact that wide regional contrasts in this respect have been recorded within short distances.

Species to be expected in the autumn shoal waters are:

Guinardia flaccida	(temperate neritic)
Thalassionema nitzschioides	(boreal neritic)
Melosira sulcata	(tychopelagic)
Chaetoceros decipiens	(boreal neritic)
Ch. compressus	(boreal neritic)
Thalassiothrix longissima	(boreal oceanic)
Rhizosolenia setigera	(boreal neritic)
Rh. hebatata var. *semispina*	(boreal oceanic)
Rh. imbricata var. *Shrubsolei*	(temperate oceanic)
Coscinodiscus centralis	(boreal oceanic)
C. excentricus	(temperate neritic)
Thalassiosira decipiens	(boreal neritic)
Th. gravida	(boreal neritic)
Sceletonema costatum	(boreal neritic)
Pleurosigma Normani	(tychopelagic)
Exuviella spp.	
Ceratium Fusus	(temperate oceanic)
C. longipes	(boreal oceanic)
C. tripos	(temperate oceanic)
Peridinium spp.	
Prorocentrum micans	(temperate neritic).

The aestival flowering of Rhizosolenia which takes place in Vineyard Sound lasts no longer than a week, in some years not more than three or four days. Other genera are known to have a much longer period of active growth in late summer or early autumn, with Sceletonema flowering for as long as six weeks in the northern part of the gulf (Gran and Braarud, 1935); nevertheless, it seems certain that active proliferation of diatoms is usually at an end over the gulf generally by the end of September or by mid-October at latest. From that date onward—by the evidence for the several years combined—the flora appears to represent essentially the transition to the winter community (p. 197), with the genus Coscinodiscus becoming dominant in the western coastal belt and on George's Bank by the end of October (as in 1933), or at least by November (e.g., in 1916), thanks largely to the decrease in the abundance of other diatoms (Figs. 11, 12).

A sparse peridinian population dominated by Ceratium also persists throughout the autumn in the coastal waters, monopolizing any areas where diatoms—e.g., Coscinodiscus— may be absent, as it then does the offshore basin generally (data from 1933), increasing somewhat in abundance during the early part of the autumn. In September and October C. Fusus is usually the most abundant species of the genus over the gulf generally, but C. lineatum was dominant (up to 460 per liter) in the very sparse flora on the western part of George's Bank in September 1933. The highest count for Prorocentrum micans was found in October when there were 2,500 cells of it per liter at a station on the eastern part of George's Bank in 1933, though it was much less numerous in the deeper waters to the north at that time.

Coccoliths also persist in moderate numbers in the basin through September (up to 2,000 per column) and October (up to 1,200,000 per column, Fig. 12).

The seasonal succession of the leading genera for representative parts of the gulf is illustrated on Fig. 13.

SEASONAL FLORAL SUCCESSION IN NEARBY LOCALITIES

The following resumé of floral successions in the Bay of Fundy, tributary to the Gulf proper and in the Woods Hole region which joins it to the west and south, based on previously published information (Fritz, 1921; Fish, 1925; Davidson, 1934; Gran and Braarud, 1935; Lillick, 1937), is added to round out the foregoing account.

BAY OF FUNDY

In the region of Passamoquoddy Bay—the only locality within the Fundian area where the phytoplankton has been studied in winter (Fritz, 1921; Davidson, 1934), the diatoms Coscinodiscus and Melosira sulcata dominate during December and January, other neritic diatoms and Dinophyceae occurring in small numbers only. This flora, which strongly resembles the winter community of the western coastal waters of the open gulf (p. 200), continues essentially unchanged until April, except for some increase in February and March in the relative importance of neritic diatoms such as Thalassionema nitzschioides and Sceletonema costatum.

The spring diatom flowering is normally in progress in Passamoquoddy Bay by the latter part of April (Fritz, 1921; Davidson, 1934; Gran and Braarud, 1935), Biddulphia aurita, Thalassiosira and Chaetoceros developing concurrently, though they reach their peaks successively. Biddulphia, which usually appears just previous to the other species,

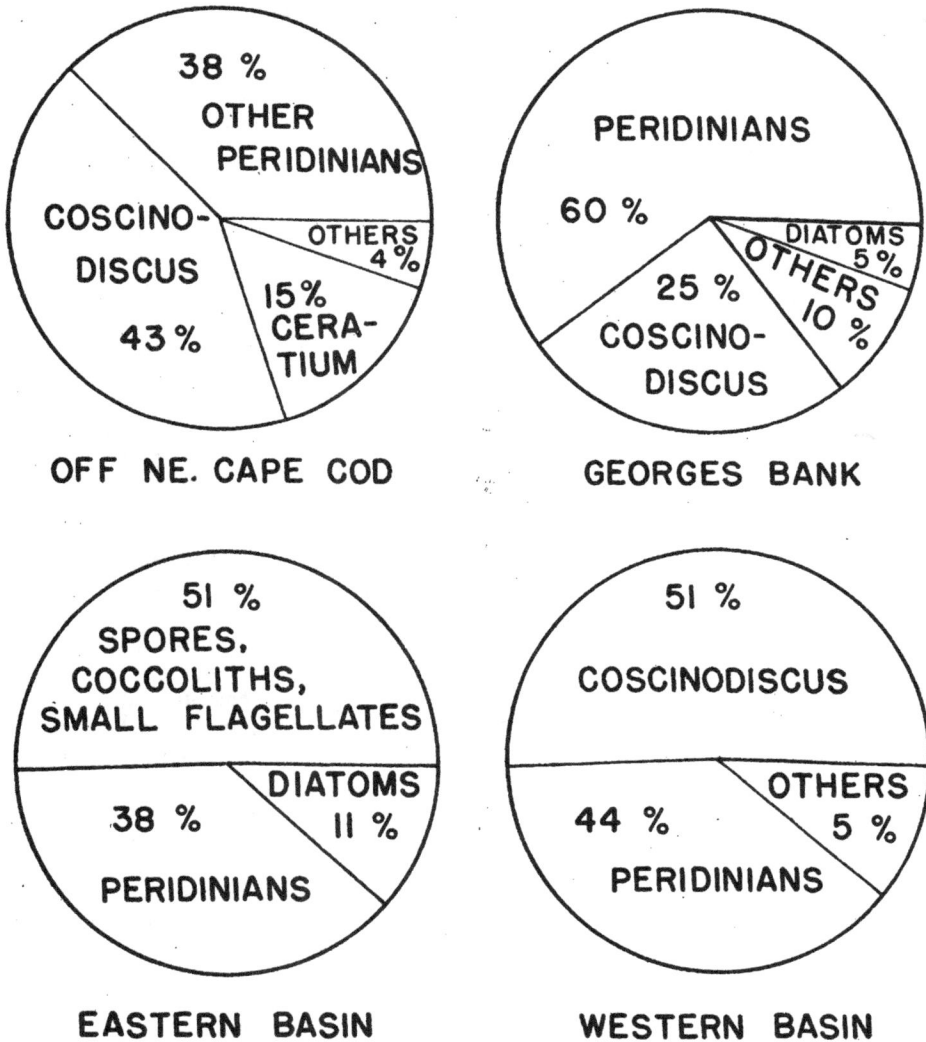

FIG. 12.   Percentage composition of representative localities in the Gulf of Maine, October 1933.

reaches its maximum abundance within a few days, and then diminishes in numbers almost as abruptly, the span of its flowering seldom lasting more than two weeks, to terminate (usually) before the end of May.   *Thalassiosira Nordenskioeldi* and its companion species *Porosira glacialis* multiply less rapidly, though to a higher level of abundance, to reach their maximum in April in some years, but not until June in others, with the period of abundance lasting for about a month. The subsequent multiplication of Chaetoceros is still slower, for this genus does not reach its peak until May in an early year, but not until July or even August in a late. As a rule the water is then dominated by *Ch. debilis*, though on occasion *Ch. compressus*, *Ch. (diadema) subsecundus* or *Ch. lacinosus* may outnumber it.[20]

The spring flora—less varied in the bay than in the gulf—is dominated by the same species in both areas. The most striking contrasts between the two localities are: (1) that the vernal flowering of diatoms takes place four to eight weeks later and generally endures longer in the bay than in the gulf; (2) that *Biddulphia aurita* is a much more important item in the flora in the former than in the latter (p. 210). After the periods of active multiplication have passed—which endure about a month for Thalassiosira and six weeks to two months for Chaetoceros—these communities abruptly disappear from the surface waters of the bay, leaving behind an impoverished flora of neritic diatoms, such as *Actinoptychus undulatus*—which is more abundant in the bay than in the gulf—*Melosira sulcata*, *Thalassiosira decipiens*, Chaetoceros spp., *Leptocylindrus danicus* and *Sceletonema costatum*. This tychopelagic and neritic association, the summer minimum of Gran and Braarud (1935, p. 339), occupies much of the bay inside Grand Manan for a short period, contrasting strongly with the Ceratium flora of late spring and summer of the open Gulf of Maine (p. 221). One of its interesting features is the abundance of the silicoflagellate *Distephanus speculum* of which Gran and Braarud (1935, p. 390) had 32,500 per liter in Passamoquoddy Bay on July 30, 1932.

In August Gran and Braarud (1935) found three distinguishable communities in the bay: (1) one along the Nova Scotian coast dominated by the neritic diatom *Ch. constrictus*, (2) another along the New Brunswick coast characterized by *Sceletonema costatum*, accompanied in less quantities by *Asterionella japonica*, *Ditylium Brightwelli* and *Coscinodiscus excentricus* and (3) their "Triquetum society" (Gran and Braarud, 1935, p. 344) of small neritic peridinians such as *Peridinium faeroense*, *P. triquetum*, *P. trochoideum* and *Goniaulax tamarensis*, which was generally distributed over the bay and present but less well developed in the northern part of the Gulf of Maine. The first of these communities Gran and Braarud (1935) class as characteristic of the drift into the bay and the second of the drift out of it; the third spreads from centers near the outlet of Digby Gut and off the mouth of the St. John's River. The oceanic species *Pontosphaera Huxleyi*, *Ceratium tripos*, *C. Fusus* and *Rhizosolenia alata*, which are so prominent in the Gulf of Maine in late summer, were also numerous in the central portion of the bay at this season in the one year of record. The presence of *Asterionella japonica* as an important member of the community in the drift out of the bay suggests a source for the flowering of that species which was encountered in the open gulf in the coastal belt between the Grand Manan Channel and Seguin Island in the summer of 1912 (p. 223).

Gran and Braarud (1935) found the Constrictus and Triquetum societies reduced almost to the vanishing point in the Bay of Fundy proper by the end of September, with the *Sceletonema* association alone persisting that late in the season in fair quantity, as it does throughout the early autumn in the northern coastal waters of the open gulf likewise. A varied diatom community in which *Chaetoceros socialis*, *Thalassiothrix longissima*, *Rhizosolenia imbricata* var. *Shrubsolei* and *Ditylium Brightwelli* are prominent persists, however,

---

[20] For graphs showing the mutual fluctuations of Biddulphia, Thalassiosira and Chaetoceros in Passamoquoddy Bay in different years see Davidson (1934).

in the Passamoquoddy region until the end of October, when it rapidly diminishes in abundance as Coscinodiscus becomes dominant, so to continue into the winter accompanied by small quantities of the various neritic diatom species which persist almost continuously throughout the year.   With the disappearance of the late summer communities from the bay, and the rise of Coscinodiscus, the flora there becomes essentially similar to that of the coastal waters of the gulf (Fig. 13).

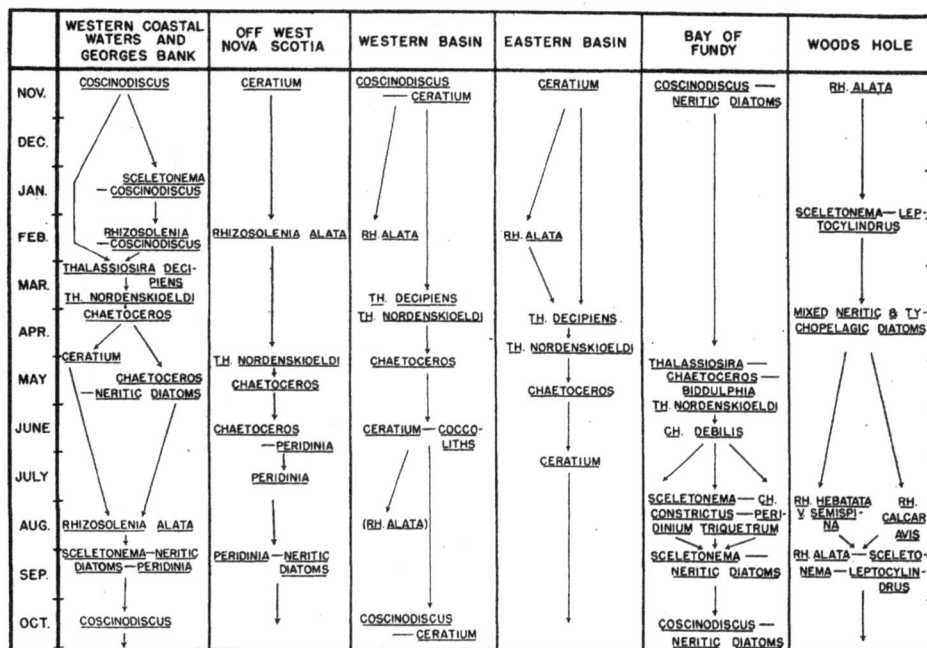

	WESTERN COASTAL WATERS AND GEORGES BANK	OFF WEST NOVA SCOTIA	WESTERN BASIN	EASTERN BASIN	BAY OF FUNDY	WOODS HOLE
NOV.	COSCINODISCUS	CERATIUM	COSCINODISCUS — CERATIUM	CERATIUM	COSCINODISCUS — NERITIC DIATOMS	RH. ALATA
DEC.						
JAN.	SCELETONEMA — COSCINODISCUS					
FEB.	RHIZOSOLENIA — COSCINODISCUS	RHIZOSOLENIA ALATA	RH. ALATA	RH. ALATA		SCELETONEMA — LEP- TOCYLINDRUS
MAR.	THALASSIOSIRA DECI- PIENS TH. NORDENSKIOELDI CHAETOCEROS		TH. DECIPIENS TH. NORDENSKIOELDI			
APR.				TH. DECIPIENS TH. NORDENSKIOELDI		MIXED NERITIC & TY- CHOPELAGIC DIATOMS
MAY	CERATIUM CHAETOCEROS — NERITIC DIATOMS	TH. NORDENSKIOELDI CHAETOCEROS	CHAETOCEROS	CHAETOCEROS	THALASSIOSIRA — CHAETOCEROS — BIDDULPHIA TH. NORDENSKIOELDI	
JUNE		CHAETOCEROS — PERIDINIA	CERATIUM — COCCO- LITHS		CH. DEBILIS	
JULY		PERIDINIA		CERATIUM		
AUG.	RHIZOSOLENIA ALATA		(RH. ALATA)		SCELETONEMA — CH. CONSTRICTUS — PERI- DINIUM TRIQUETRUM	RH. HEBATATA RH. V SEMISPI- CALCAR NA AVIS
SEP.	SCELETONEMA — NERITIC DIATOMS — PERIDINIA	PERIDINIA — NERITIC DIATOMS			SCELETONEMA — NERITIC DIATOMS	RH. ALATA — SCELETO- NEMA — LEPTOCYLIN- DRUS
OCT.	COSCINODISCUS		COSCINODISCUS — CERATIUM		COSCINODISCUS — NERITIC DIATOMS	

FIG. 13.   Seasonal succession of dominants in representative parts of the Gulf of Maine
(derived from all sources).

WOODS HOLE REGION

Although the Woods Hole region lies so close to the Gulf of Maine, the phytoplanktonic cycle is very different there, for the main flowering of diatoms has been found to take place in the winter instead of in spring in two of the three years of record, both in Vineyard Sound and in Buzzards Bay (Fish, 1925, Lillick, 1937), commencing in mid-November, reaching its maximum in December and continuing until March when the flora suddenly diminishes in abundance (Fig. 13).   Dominance in this winter maximum is shared by a large number of important species, *Rhizosolenia alata* usually leading in abundance early in the season, to be replaced later by *Leptocylindricus danicus* and by *Sceletonema costatum*. Other species that may be codominant are *Ditylium Brightwelli, Thalassionema nitzschioides, Rhizosolenia setigera, Rh. imbricata* var. *Shrubsolei* and *Chaetoceros socialis* (all of them neritic) and the oceanic forms *Nitzschia seriata* and *Ch. decipiens*.   In one year (i.e., 1935–1936) it appears that no definite maximum developed either in winter or in spring (Lillick, 1937).

A sparse flora consisting of neritic or tychopelagic species prevails in the Woods Hole region throughout the later spring and early summer, including *Guinardia flaccida*, *L. danicus*, *T. nitzschioides*, *Prorocentrum micans*, *P. scutellum*, *S. costatum*, *Corethron hystix*, *N. seriata*, species of Chaetoceros and others, but with wide variation.   A second brief flowering dominated by Rhizosolenia occurs in mid- or late summer.   In the four summers 1935–1938, during which the writer has observed this maximum, the dominant species was either *Rh. semispina* or *Rh. calcar-avis*, both occurring in great abundance; but *Rh. imbricata*, var. *Shrubsolei* and *Rh. setigera* were also present in numbers sufficient for either of these to dominate, if hydrographic conditions were such as to prevent a flowering of *Rh. semispina* (Fish, 1925).   Small amounts of *G. flaccida*, *S. costatum* and peridinians likewise contribute to this flora.   It has been suggested that this flowering represents a series of local swarms; actually, however, the same association of species is generally distributed over the Woods Hole area at the time.   In the years of record this flowering was at its peak late in July or early in August and lasted for little more than a week.   Diatoms then decreased so abruptly in abundance that only a very scanty population was left, consisting of *Rh. alata*, *S. costatum* and *L. danicus* accompanied by a variety of other species in minimal numbers, a situation which continues until the winter maximum again develops.

The Woods Hole flora contrasts further with that of the Gulf of Maine in the occurrence of a much larger number of small pigmented flagellates which form a significant portion of the phytoplankton there, especially during the warm months of the year (Lackey, 1936).

The Woods Hole area and the waters thence to Long Island Sound agree in seasonal schedule with the Mediterranean and with the Adriatic Seas more closely than with the Gulf of Maine, as Fish (1925) has pointed out.   In fact it is only during the late summer flowering of Rhizosolenia that the Woods Hole flora resembles that of the Gulf, and even then different species dominate in the two areas.

COMPARISON WITH OTHER BOREAL REGIONS

It has already been noted by Bigelow (1926), by Føyn (1929) and by others, that the planktonic cycle of the Gulf of Maine parallels that of the North Sea and of the waters of southern Norway, both in time-sequence and in dominant species.   The northern Norway Sea (Lofoten) the Irish Sea, the Clyde Sea area, the English Channel and the continental shelf off Sydney, Australia are also similar as to cycle, though not so as to species (Herdman, 1918; Herdman, Scott and Dakin, 1910; Marshall and Orr, 1926, 1929; Harvey, Cooper, Lebour, and Russell, 1935; Dakin, 1934; Dakin and Colefax, 1935).   The phytoplankton cycle of the Gulf of Maine is even similar to that of certain areas of the South Atlantic (Hart, 1935), with the same or related genera dominant at comparable stages in many cases. Especially interesting in connection with the gulf is the survey Braarud and Bursa (1939) conducted in the Oslo Fjord during the same year (1933–1934) in which our general surveys of the gulf were made, for the cycle in the outer part of the fjord is typical of Norwegian coastal waters in general.   In June they report an abundant diatom flora dominated by *Sceletonema costatum*, but with species of Ceratium also conspicuous.   During the summer

Rhizosolenia alata, Ceratium and species of Peridinium were abundant, but by October these had been replaced by a community relatively poor in diatoms and ceratia. In December the flora consisted in the main of a weak development of Sceletonema and of the coccolithophorid *Calsiosolenia Grani*. A Thalassiosira-Porosira-Sceletonema flowering makes up the spring maximum in this region early in March, but this stage was missed by the collections. A Chaetoceros plankton was found in April, and Sceletonema in May, accompanied by *Pontosphaera Huxleyi*. The similarity between this cycle and that of the coastal waters of the Gulf of Maine (Fig. 13) is apparent. The outstanding differences are the greater prominence of Sceletonema in Norwegian waters—in European waters in general—the insignificance of Coscinodiscus in the winter flora there and a difference in the flowering dates.

The seasonal cycle of the Bay of Fundy, especially of Passamoquoddy Bay (Davidson, 1934), parallels that along the northwest coast of America from Friday Harbor, Washington, to the Aleutian Islands and Scotch Cap, Alaska (Cupp, 1937), except that the various stages follow about four to eight weeks later in the latter area. Although the specific composition of the flora of the two regions is not always the same, especially during the early spring, the same dominant genera are common to both; the chief divergent feature noted by Cupp (1937) is the greater abundance in late summer and autumn of *Asterionella japonica* in Alaskan waters than in the Bay of Fundy, a difference which is minimized in the light of recent work in the bay (Gran and Braarud, 1935).

The outstanding fact to be gleaned from these comparisons is that boreal waters, widely separated though they may be as to latitude and longitude, or by intervening land masses which prevent any direct connection by ocean currents, may bear strong resemblance one to another, both as to the cyclic production of cells, and as to the generic and even specific composition of the flora.

SUMMARY

The flora of the Gulf of Maine proper is not only scanty in early winter but also the least varied then, dominated as a rule either by Coscinodiscus, by Ceratium or by small species of other peridinian genera, the relative importance of which varies considerably from year to year, as outlined above (p. 197). Coscinodiscus reaches its maximum abundance as a whole in midwinter, with maximum frequency up to 1,600 per liter. *Rhizosolenia alata* has also been found flowering locally in moderate abundance (up to 1,300 per liter) at that season on both sides of the gulf, and increases have been recorded in late December or January for *Thalassionema nitzschioides*, for *Sceletonema costatum* and for others of the species listed on p. 203. Coincidently, however, the Dinophyceae decrease. The chief variation from year to year at this season is in the relative proportions of Coscinodiscus as compared to peridinians, and in the local development of associations of unusual composition as described on p. 205.

The outstanding alteration from winter to spring is the vernal outburst of diatoms, accompanied by a still further decrease in the number of peridinians, actual as well as relative. This outburst results chiefly from the rapid multiplication of Thalassiosira, a genus very scarce or absent in early winter, but appearing in significant numbers in late January or in February. In the open gulf *Th. Nordenskioeldi* is usually the dominant species; toward the north, however, *Th. decipiens* may equal or outnumber it, as seems to be usual in the Bay of Fundy, and in some years *Th. gravida* also is co-dominant. *Porosira glacialis* may likewise be codominant in the bay, though not so recorded in the gulf proper. Other species characteristic of the Thalassiosira flora of spring are listed on p. 209.

The vernal flowering of Thalassiosira develops earliest in the western coastal belt, at a date varying between late February and early April in different years; soon afterward (March–mid-April) in the coastal and bank waters west and south of Nova Scotia. It is possible that early flowerings of Thalassiosira also develop on the western part of George's Bank, but the evidence is not clear in this respect. The areas occupied by the flowering of Thalassiosira extend northward along shore in both sides of the gulf to join across the mouth of the Bay of Fundy; as well as southward and offshore into the eastern side of the open basin. They also involve the western basin, though it is not known whether Thalassiosira ever dominates as strongly there (p. 208), or on the eastern part of George's Bank, as it does in the coastal waters of the gulf generally. Active multiplication of Thalassiosira is briefest —2–4 weeks—in the eastern and southeastern parts of the gulf generally, and on George's Bank, longest in the northern coastal belt (8 to 12 or more weeks) and in Passamoquoddy Bay, tributary to the Bay of Fundy. The western coastal belt is intermediate in this respect, the flowering enduring there for 4–7 weeks (average about 5 weeks). ·

Coincident with the vernal flowering of Thalassiosira, the eastern side of the gulf is invaded by a weak population of the cold-water peridinian *Ceratium arctica* which, however, disappears after April. After Thalassiosira has reached its peak it declines so abruptly in abundance over the open gulf that within a few days it falls to an insignificant rank in the flora. The rise and fall of *Th. Nordenskioeldi* is closely correlated with temperature, its optimum being below 5° C.

The eclipse of Thalassiosira is yearly accompanied throughout the area by an active flowering of the genus Chaetoceros, except perhaps in the eastern basin for which the evidence is conflicting. This flowering reaches its climax at about the time Thalassiosira approaches its minimum, i.e., in some time in April over the deeper parts of the gulf generally, but not until the end of May or even later in the northern coastal belt, where Thalassiosira persists longest. In any given part of the open gulf the alternation in dominance between the two genera usually takes place within 10 days, with the regional expansion of Chaetoceros correspondingly paralleling that characteristic of Thalassiosira earlier in the season. At their respective peaks of abundance Chaetoceros ranks second to Thalassiosira in the gulf (see Table, p. 213). During its vernal peak 15 species of Chaetoceros have been identified (Table I), most of them, however, being in small numbers. The most frequently dominant is *Ch. debilis*, as may be *Ch. constrictus* also in some years. Other species have also been found dominant locally as described on p. 214. Present indications are that the normal duration of active vernal flowering for Chaetoceros is 4–6 weeks over the gulf generally, i.e., until late April or some time in May. But the fact that, although its period of rapid multiplication endured for 3 weeks only near Cape Ann in 1938, this continued throughout the early summer (i.e., for more than 8 weeks) in the western coastal belt in 1914 and again in 1934, shows that considerable variation is to be expected in this respect from year to year, as is also the case in the Bay of Fundy.

In general, Chaetoceros declines as abruptly in abundance as does Thalassiosira in the open gulf, once its peak of abundance is passed—for illustration of this see p. 213; in some years, however, it may persist in moderate numbers throughout the entire summer at scattered localities in the western and northern coastal belts. In 1920 the eclipse of Chaetoceros was succeeded in Massachusetts Bay by a brief but intense flowering of Phaeocystis, but this is the only such record for the gulf; hence it is doubtful whether it is a regular event in the seasonal cycle, or whether the case in point was a sporadic occurrence.

Diatoms of one or another of the species listed on p. 221 continue to dominate the flora throughout the summer on George's Bank, locally, however, and sporadically, and are sometimes outnumbered there by Pontosphaera. They may also dominate an impoverished flora all summer in the northern coastal belt in some years. In other years, however, the eclipse of Chaetoceros leaves this latter belt barren of diatoms and dominated by a sparse dinophycean flora, as it does the coastal belt farther south and the bowl of the gulf as a whole, at a date varying locally and with the years between the end of May and the end of June or beginning of July. In the eastern coastal belt the situation is complex, requiring more extensive data for its clarification. The peridinian flora of summer is dominated by species of Ceratium among which *C. tripos*, *C. longipes*, *C. bucephalum* and *C. Fusus* are the most important, accompanied by the other peridinians, diatoms and coccolithophorids listed on p. 218. Earlier accounts of the mutual fluctuations in abundance of *C. longipes* and *C. tripos* (Bigelow, 1926) are confirmed by the data for more recent years.

Chief alteration of the flora during the late summer is such as results from the local development of second flowerings of diatoms, Rhizosolenia being chiefly responsible for these in the shoaler waters including George's Bank. For details as to abundance at this season of individual species of this genus see p. 222. Chaetoceros may also have a distinct second flowering close inshore in the northern coastal belt, and *Guinardia flaccida* is an important component of the late summer flowering, especially on George's Bank. *Sceletonema costatum* also has its maximum in late summer (up to 700,000 cells per liter) when it may dominate locally inshore (as near Mt. Desert Island, off Nova Scotia and in Nantucket Sound) in August, and in Massachusetts Bay in September. *Thalassiothrix longissima* similarly dominated a rich flora on the western part of George's Bank late in July 1916, and *Asterionella japonica*, a species widespread in summer in the Bay of Fundy, but usually scarce or wanting in the open gulf, flowered in great abundance in the northern coastal belt in August of 1912. Peridinians are overshadowed in the shoaler parts of the gulf during the late summer flowerings of diatoms, but the same species persist there as in early summer, and in undiminished quantity, besides others as listed on p. 221. Peridinians as a group dominate in the bowl of the gulf generally throughout the summer except on the western side, to which the Rhizosolenia flowering may extend from the coastal belt, as described on p. 226. It is at this season, also, that coccoliths are at their maximum—notably *Pontosphaera Huxleyi*, a population of nearly 300,000 of which per liter was recorded on George's Bank and in the neighboring basin in July 1933 (Braarud, 1934). The only record of unicellular plants of tropical origin in the gulf is of occasional *Ceratium macroceros*.

Sceletonema has been found flowering during the first half of September in the northern gulf and until late in that month in Massachusetts Bay. Guinardia also was in great abundance near Seguin Island and on the eastern part of George's Bank in early September 1933, when other diatoms also in wide variety were recorded at one station or another. In fact it is at this season that the diatom flora may be the most varied (for list of species see p. 227) and the regional contrasts widest in its qualitative composition. Active proliferation of diatoms is generally ended, however, in most parts of the gulf by the end of September or by early October, even in localities where it persists the latest. From that time onward the flora gradually assumes the early winter state, the quantitative impoverishment of diatoms as a group leading once more to a sparse flora dominated by Coscinodiscus, by peridinians, or by the two in combination.

The seasonal cycle is essentially the same in the Bay of Fundy as in the open gulf except that Biddulphia is a more important factor in the vernal flowering, that the vernal peaks of abundance for Thalassiosira and Chaetoceros develop later in the bay and endure longer there, and that tychopelagic and neritic diatoms of the species named on p. 230 are relatively more important there in summer, and Peridinians relatively less so.

In the Woods Hole region, however, the maximum flowering takes place three months earlier in the season than in the gulf (i.e., mid-November to early March), and is chiefly of species that are of little importance in the winter-spring flora of the latter, whereas Thalassiosira—the leading genus in the gulf—is negligible westward from Cape Cod.

The records for 1933–1934 corroborate earlier observations that the cycle of planktonic plant production in the Gulf of Maine closely parallels that of boreal waters generally on the opposite side of the North Atlantic in time-sequence, though different associations of species may dominate the peaks of abundance in different localities, under different environmental conditions.

BIBLIOGRAPHY

BAILEY, J. W.
  1854 Notes on new species and localities of microscopical organisms. Smithsonian Contrib. Knowl., Vol. 7, pp. 4–15, 4 textfigs., 1 pl.

BIGELOW, H. B.
  1914 Oceanography and plankton of Massachusetts Bay and adjacent waters, November, 1912–May, 1913. Bull. Mus. Comp. Zool., Vol. 58, No. 10, pp. 385–419, 7 textfigs., 1 pl.
  1917 Explorations of the coast water between Cape Cod and Halifax in 1914 and 1915, by the U. S. Fisheries schooner Grampus. Oceanography and plankton. Bull. Mus. Comp. Zool., Vol. 61, No. 8, pp. 161–357, 100 textfigs., 2 pls.
  1926 Plankton of the offshore waters of the Gulf of Maine. Bull. U. S. Bureau of Fisheries, Vol. 40, Pt. I, pp. 1–509, 134 textfigs.
  1927 Physical oceanography of the Gulf of Maine. Bull. U. S. Bureau of Fisheries, Vol. 40, Pt. II, pp. 511–1027, 207 textfigs.

BIGELOW, H. B., LILLICK, LOIS C., AND SEARS, MARY
  1940 Phytoplankton and planktonic protozoa of the offshore waters of the Gulf of Maine. Part I. Numerical distribution. Trans. Amer. Philos. Soc., Vol. 31, Pt. III, pp. 149–191, 10 textfigs.

BRAARUD, Trygue
  1934 A note on the phytoplankton of the Gulf of Maine in the summer of 1933. Biol. Bull., Vol. 67, pp. 76–82.

BRAARUD, T., AND BURSA, A.
  1939 The phytoplankton of the Oslo Fjord, 1933–1934. Hvalrådets Skrift. Norske Vidensk.-Akademi i Oslo, No. 19, pp. 1–63, 9 textfigs.

BURKHOLDER, P. R.
  1933 A study of the Phytoplankton of Frenchmans Bay and Penobscot Bay, Maine. Internat. Rev. d. ges. Hydrobiol. u. Hydrogr., Vol. 28, pp. 262–284, 6 textfigs.

CUPP, E. S.
  1937 Seasonal distribution and occurrence of marine diatoms and dinoflagellates at Scotch Cap, Alaska. Bull. Scripps Inst. Oceanogr. La Jolla, Cal. Tech. Ser., Vol. 4, pp. 71–100.

DAKIN, W. J.
  1934 The plankton calendar of the continental shelf of the Pacific coast of Australia at Sydney, compared with that of the Irish Sea. James Johnstone Memorial Volume, pp. 164–175. Univ. Press of Liverpool.

DAKIN, WILLIAM J., AND COLEFAX, ALLEN N.
  1935 Observations on the seasonal changes in temperature, salinity, phosphates, and nitrate nitrogen and oxygen of the ocean waters of the continental shelf off New South Wales and the relationship to plankton production. Proc. Linn. Soc., New South Wales, Vol. 40, pp. 303–314, 11 textfigs., 1 pl.

DAVIDSON, V. M.
  1934 Fluctuations in the abundance of planktonic diatoms in the Passamoquoddy region, New Brunswick, from 1924 to 1931. Contrib. Canadian Biol. and Fish., N. S., Vol. 8, No. 28, pp. 357–407, 33 textfigs.

FISH, C. J.
  1925 Seasonal distribution of the plankton of the Woods Hole region. Bull. U. S. Bureau of Fisheries, Vol. 41, pp. 91–179, 81 textfigs.

FØYN, BIRGITHE RUUD
  1929 Investigation of the phytoplankton at Lofoten, March–April, 1922–1927. Skrift. Norske Vidensk.-Akademi i Oslo, 1928, Matem.-Naturvidenskap. Klasse, Vol. 1, pp. 1–71, 15 textfigs.

FRITZ, CLARA W.
  1921 Plankton Diatoms, their Distribution and Bathymetric Range in St. Andrew's Waters. Contrib. Canadian Biol., 1918–1920, Dept. Naval Service, pp. 49–62, 3 pls.

GRAN, H. H.
  1904 Die Diatomen der arktischen Meere. 1 Theil: Die Diatomen des Planktons. Fauna Arctica, Vol. 3, pp. 509–554, Jena.
  1933 Studies on the biology and chemistry of the Gulf of Maine. II. Distribution of phytoplankton in August, 1932. Biol. Bull., Vol. 64, No. 2, pp. 159–182.

GRAN, H. H., AND BRAARUD, Trygve
  1935 A Quantitative Study of the Phytoplankton in the Bay of Fundy and the Gulf of Maine (including Observations on Hydrography, Chemistry and Turbidity). Jour. Biol. Board of Canada, Vol. 1, pp. 279–467, 69 textfigs.

HART, T. J.
  1935 On the Phytoplankton of the South West Atlantic and the Bellingshausen Sea, 1929–1931. Discovery Reports, Vol. 8, pp. 1–268, 84 textfigs., 54 tables.

HARVEY, H. W., COOPER, L. H. N., LEBOUR, M. V., AND RUSSELL, F. S.
  1935 Plankton Production and its Control. Jour. Mar. Biol. Assoc., N. S., Vol. 20, pp. 407–441, 16 textfigs.

HERDMAN, W. A.
  1918 Spolia Runiana. III. The Distribution of certain Diatoms and Copepoda, throughout the year, in the Irish Sea. Jour. Linn. Soc., Botany, Vol. 44, pp. 173–204, 21 textfigs. (Reprinted in Jour. Linn. Soc., Zoology, Vol. 34, pp. 95–126, 21 textfigs.)

HERDMAN, W. A., SCOTT, ANDREW, AND DAKIN, W. J.
  1910 An intensive study of the marine plankton around the south end of the Isle of Man. Part III. Proc. and Trans. Biol. Soc., Liverpool, Vol. 24, pp. 255–359, 21 textfigs.

LACKEY, J. B.
  1936 Occurrence and distribution of the marine protozoan species in the Woods Hole area. Biol. Bull., Vol. 70, pp. 264–278.

LILLICK, LOIS C.
  1937 Seasonal studies of the phytoplankton off Woods Hole, Massachusetts. Biol. Bull., Vol. 73, No. 3, pp. 488–503, 3 textfigs.
  1938 Preliminary report of the phytoplankton of the Gulf of Maine. Amer. Midl. Nat., Vol. 20, pp. 624–640, 1 textfig.

MARSHALL, S. M., AND ORR, A. P.
  1926 The Relation of the Plankton to some Chemical and Physical Factors in the Clyde Sea Area. Jour. Mar. Biol. Assoc., N. S., Vol. 14, pp. 837–868, 9 textfigs., 10 pls.
  1929 A Study of the Spring Diatom Increase in Loch Striven. Jour. Mar. Biol. Assoc., N. S., Vol. 16, pp. 853–878, 15 textfigs.

SPARROW, F. K., JR.
  1937 The occurrence of saprophytic fungi in marine muds. Biol. Bull., Vol. 73, pp. 242–248.

www.ingramcontent.com/pod-product-compliance
Lightning Source LLC
Chambersburg PA
CBHW081335190326
41458CB00018B/6011